Manhattan Water-Bound

New York City History and Culture
Jay Kaplan, Series Editor

Manhattan Water-Bound

Manhattan's Waterfront from the
Seventeenth Century to the Present

SECOND EDITION

Ann L. Buttenwieser

With a Foreword by Robert A. M. Stern

Syracuse University Press

This book is published with the assistance
of a grant from the John Ben Snow Foundation.

The first edition was published in 1987 by New York University Press.

The paper used in this publication meets the minimum requirements of
American National Standard for Information Sciences—Permanence of
Paper for Printed Library Materials, ANSI Z39.48-1984.

Library of Congress Cataloging-in-Publication Data
Buttenwieser, Ann L., 1935–
    Manhattan water-bound : Manhattan's waterfront from the
seventeenth century to the present / Ann L. Buttenwieser ; with a
foreword by Robert A. M. Stern. — 2nd ed.
        p.   cm. — (New York City history and culture)
    Originally published: 1st ed. New York : New York University
Press, 1987.
    Includes bibliographical references and index.
    ISBN 0-8156-2801-3 (pbk. : alk. paper)
    1. City planning—New York (State)—New York—History.
2. Waterfronts—New York (State)—New York—History.   3. Manhattan
(New York, N.Y.)—History.   I. Title.   II. Series.
HT168.N5B88   1999
307.1'2'097471—dc21   98-55154

Manufactured in the United States of America

For (in chronological order) Indy, Ashley, Hallie,
Nicholas, Else, Sam, Liberty, Julia, and Zach

Ann L. Buttenwieser developed an interest in waterfronts while growing up in Annapolis, Maryland. She has pursued her interest planning the waterfronts that are under the jurisdiction of several New York City agencies, including the Department of Planning, the Department of Parks & Recreation, and the Economic Development Corporation. In the 1980s, she was Deputy Director of the West Side Task Force, seeking solutions for Manhattan's Hudson River waterfront.

Contents

Contents

Contents

Figures

Figures

Maps

Tables

Agency and Literary Abbreviations

Assessments	Annual Record of Assessed Valuation of Real Estate
Assistant Aldermen	New York City Board of Assistant Aldermen
BPM	Borough President of Manhattan
BSA	Board of Standards and Appeals
City Clerk	City Clerk Approved Papers, New York City Municipal Archives
Conveyances	New York County Register's Office, Re-Indexed Conveyances of the Blocks and Lots of the City of New York . . .
DD	New York City, Department of Docks also called: Docks and Ferries; Marine and Aviation; Ports and Terminals; and Ports International Trade & Commerce

Docks Meetings	*Public Meetings of the Department of Docks to Hear Persons Interested in Improving the Waterfront*
EDC	New York City Economic Development Corporation, formerly the Public Development Corporation
EIS, DEIS and FEIS	Environmental Impact Statement; Draft and Final Environmental Impact Statement
EPA	Environmental Protection Agency
ESC News	*East Side Chamber News*
HA Papers	New York City Housing Authority Papers, LaGuardia Community College Archives
Henry Street	Henry Street Studies, *A Dutchman's Farm: 301 Years At Corlears Hook.* 1939. (New York: Henry Street Studies)
Henry Street Collection	Henry Street Settlement Collection, Social Welfare History Archives, University of Minnesota
ISTEA	Intermodal Surface Transportation Efficiency Act
Mayors Papers	New York City Municipal Archives Collection, organized by names of the city's mayors
New York State DOT	New York State Department of Transportation
N.Y.S., *N.Y. Harbor Commission*	*Report of the Harbor Commission of 1856 and 1857*

NYT	*New York Times*
OPRHP	New York State Office of Parks, Recreation and Historic Preservation
Port Authority	The Port of New York Authority, now called the Port Authority of New York and New Jersey
PDC	Public Development Corporation
PWA	Federal Public Works Authority
RFP	Request for Proposals
RPA	Regional Plan Association
RRBG	*Real Estate Record and Builders' Guide*
SI Papers	Stanley Isaacs, Manuscript Collection, New York Public Library
ULURP	Uniform Land Use Review Procedure
UDC	Urban Development Corporation, now the Empire State Development Corporation
WL	Womens' League for the Protection of Riverside Park
WPA	Writers' Program of the Work Projects Administration; originally called Works Progress Administration

Foreword

Robert A. M. Stern

New York is what it is because of its harbor. The harbor was
the reason the exploring Dutch stayed and prospered. And the
English after them. In fact, the harbor was central to the city
until about the time of the Civil War when, as land-bound
transportation came to rival that on water, and then to dominate it, the waterfront became less integral to daily life in
Manhattan. By the end of the nineteenth century, with the
exception of transatlantic travel, the principal activities of the
port were shifting off island to Brooklyn and to New Jersey.
With the decline of water travel, Manhattan turned increasingly in on itself, looking away from dilapidated piers and
water-hugging industrial areas—its manufactories, its markets
and its abattoirs—to a glorious new interior asset, Central
Park, which became and remains the focus of public life.

In the post–Civil War era, the Department of Docks was
founded as the result of pressure from leading citizens and
business interests who wanted municipal officialdom to step
in and plan for an area whose haphazard private development

not only threatened what was then left of the harbor's economic life but also that of all the abutting upland areas. New docks were built, but the grand plans of the 1870s were never realized. In fact, not much happened until the 1930s when the waterfront was seen as the best place to put highways, a mentality that lasted until the proposal of the much misunderstood Westway, which ended in a debacle but broke the prevailing model of thought. Until Manhattanites became exercised over Westway, there was no strong public constituency for a planned waterfront, much less one that favored people over cars. Ultimately defeated, Westway re-woke Manhattan to the virtues of its long-neglected asset, transforming the waterfront into a hotly contested piece of Manhattan real estate.

When Ann Buttenwieser's book was first published in 1987, it represented that rare and happy coincidence of a propitiously timed scholarly work. Coming on the heels of the Westway debacle, it provided just what was most needed: a comprehensive, historically based overview of the situation. By presenting the history of failed vision in a meticulously researched and judiciously stated way, Buttenwieser gave us an education in the hyper-reality that is the ongoing political life of our city. Buttenwieser did much to clear the stale air of Westway that tainted most thinking about the waterfront and spurred us all to get on with the job of riverfront reclamation. As such, *Manhattan Water-Bound* is one of the rare books of urban history that made history happen. It made us see the waterfront for what it could be, an asset that was literally visible to all but almost totally overlooked.

In the decade that has passed since the initial publication of Buttenwieser's landmark study, interest in the subject has increased and, as importantly, action has actually been taken. How appropriate it is that a new edition should be published with additional coverage bringing the story up-to-date; and how nice it is to know that the book will be read or read again at a moment when ferries once again ply our rivers and new quai-side promenades are packed with people.

The waterside is Manhattan's great urban frontier. Despite the failure of the Westway project to gain a powerful enough constituency, large-scale projects like Riverside South and relatively small projects like the new Chelsea Piers are potent reminders that the waterfront offers one of the best places to make grand plans that can benefit everyday life—and to get something done. New York has certainly done its best over the years to try to kill its waterfront. Though some blows have landed hard, the potential for rejuvenation, not to mention reinvention, remains. Indeed, who would have guessed that at the end of the twentieth century the water would actually be getting cleaner and closer to where New Yorkers work and live, seek their recreation and even catch edible fish? No little thanks to this small book full of big ideas.

Acknowledgments

This edition of *Manhattan Water-Bound* is accompanied by gratitude and affection for all who encouraged, opened collections, sent material, critiqued, answered queries, suggested new lines of inquiry, and along the way had infinite patience.

The maps and schematic drawings were perseveringly prepared by Eric Meuller. Bill Woods supplied us with a finally up-to-date layout of the waterfront. Becket Logan once again worked magic on the photographs. And Francis Grunow used his excellent research skills to seek out relevant testimony.

Thanks are also due to Larry Buttenwieser, Page Samson, and Jeannette Rausch, who offered their perceptive comments; to Marcia Reiss, who lent me her files; and to Carl Weisbrod, Michael O'Connor, and my family, who gave me time to write.

Manhattan Water-Bound

Introduction

The waterfronts of North American cities have come alive! From Davenport to New Orleans, New York to Seattle, new-towns-in-town have risen on the reinforced foundations of decaying piers; shopping malls have replaced abandoned warehouses; parks have greened once-concrete shorelines; and ferries, tour boats, and gambling and pleasure craft are using waterways depleted of bulk cargo ships. Although the activities have changed, the names—Harborplace, The Anchorage, Portside Park, and Seaport Village—attempt to recall a seafaring heritage. The resurgence represents but one of many periods of optimism occurring after years of declining facilities and receipts.

This revitalization is happening in many North American port cities. By virtue of its complex geography, its monumental commercial importance, and the length of its history, however, New York City's waterfront is most interesting and most instructive. As the growth and decline of the perimeter of Manhattan Island is traced through case studies of the Dock

Introduction

Department Plan of 1870 and the Chelsea Piers, Riverside Park and Stuyvesant Cove, the East River Drive and Westway, and Vladeck Houses and Riverside South, several themes keep appearing. These themes—the importance of individual and group pressure, the ad hoc nature of government intervention, and the unforeseen consequences of changing technology— are echoes of the experiences of other waterfront cities, and highlight some of the reasons why New York City's shores developed as they did. The recurring motifs also place the constraints on contemporary revitalization efforts in a historic perspective.

Although this development is basically a history of physical expansion—of land, wood, and concrete—many individuals and groups influenced the forms it took. They were as diverse as the shoreline itself: tugboat owners; the city's prime realtors; members of the Greater New York Taxpayers' Association; George McClellan, a Civil War engineer and presidential candidate; Helen Carver Kerr, Upper West Side matron; Robert Moses, master builder; Helen Hall, Lower East Side settlement house director; Al Butzel, environmental attorney; Richard Kahan, imaginative public servant; and Roland Betts, film financier, all tried to transform a portion of the city's waterfront. In every period, people had grand visions, but the shoreline ultimately developed through compromise and through piecemeal reforms.

Government in many guises has also tried to manage and shape the shoreline. In the past when New York property owners failed to create the structures essential to maritime commerce, the city's Common Council or its Dock Department intervened. When change again proved to be elusive, other jurisdictions, such as the state harbor commissions or the U.S. Army Corps of Engineers, were brought in, and new bodies with bonding powers, such as the bistate Port of New

York Authority, were created. In addition, federal highway and housing dollars have historically and continue today to profoundly affect the configuration of the shore. In all of these endeavors, government emerged as regulator and temporary financier. Not until the late twentieth century did it assume the role of planner, thus adding to the uneven nature of waterfront change.

The effect of technology on the form, the use, and the value of the riversides crops up again and again throughout the island's history, as does the desire to reach the water for recreation. Mechanical pumping systems, for example, expedited the creation of new land by facilitating bulkhead improvements and pouring foundations for new real estate ventures. The invention of the steam turbine engine and containerization, on the other hand, suddenly made vast amounts of shoreline useless. New waste treatment methods, in turn, cleaned up the rivers and created a demand for beaches and boating. Instead of anticipating these changes, the city was always trying to catch up, causing whatever plans existed to be further compromised.

The search for a better urban shoreline is an old one. On 2 December 1871, the *Real Estate Record and Builders' Guide* warned the city not to be parsimonious when redeveloping its waterfront. "The dock system should be developed in its utmost perfection," the author wrote. The system will be a "monument to the largeness of view and forethought of the men who planned [it] and will endure as long as the city endures." Today, with most of the dock system in transformation to meet perceived twenty-first-century needs, we are still looking for the perfect solution.

Table 1. Waterfront Time Line

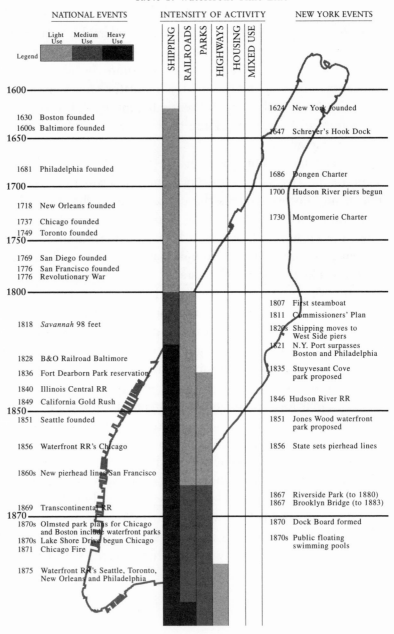

	NATIONAL EVENTS			INTENSITY OF ACTIVITY							NEW YORK EVENTS
		Light Use / Medium Use / Heavy Use (Legend)		SHIPPING	RAILROADS	PARKS	HIGHWAYS	HOUSING	MIXED USE		
1600											
	1630 Boston founded										1624 New York founded
	1600s Baltimore founded										1647 Schreyer's Hook Dock
1650											
	1681 Philadelphia founded										1686 Dongen Charter
1700											1700 Hudson River piers begun
	1718 New Orleans founded										
	1737 Chicago founded										1730 Montgomerie Charter
1750	1749 Toronto founded										
	1769 San Diego founded										
	1776 San Francisco founded										
	1776 Revolutionary War										
1800											
											1807 First steamboat
											1811 Commissioners' Plan
	1818 *Savannah* 98 feet										1820s Shipping moves to West Side piers
											1821 N.Y. Port surpasses Boston and Philadelphia
	1828 B&O Railroad Baltimore										1835 Stuyvesant Cove park proposed
	1836 Fort Dearborn Park reservation										
	1840 Illinois Central RR										
	1849 California Gold Rush										1846 Hudson River RR
1850											
	1851 Seattle founded										1851 Jones Wood waterfront park proposed
	1856 Waterfront RR's Chicago										1856 State sets pierhead lines
	1860s New pierhead lines San Francisco										
											1867 Riverside Park (to 1880)
	1869 Transcontinental RR										1867 Brooklyn Bridge (to 1883)
1870											
	1870s Olmsted park plans for Chicago and Boston include waterfront parks										1870 Dock Board formed
	1870s Lake Shore Drive begun Chicago										1870s Public floating swimming pools
	1871 Chicago Fire										
	1875 Waterfront RR's Seattle, Toronto, New Orleans and Philadelphia										

Table 1 (cont.)

NATIONAL EVENTS	INTENSITY OF ACTIVITY						NEW YORK EVENTS
	SHIPPING	RAILROADS	PARKS	HIGHWAYS	HOUSING	MIXED USE	

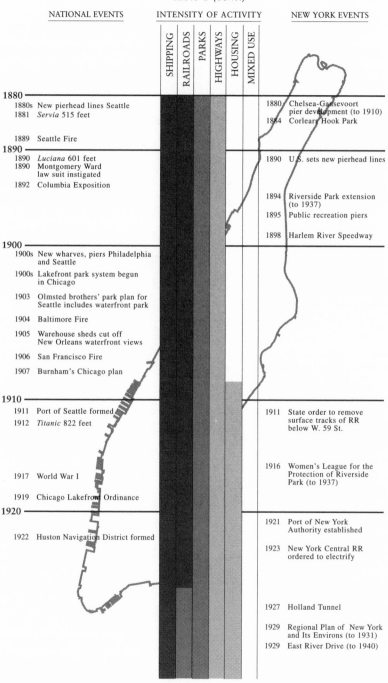

1880

1880s New pierhead lines Seattle
1881 *Servia* 515 feet

1889 Seattle Fire

1890

1890 *Luciana* 601 feet
1890 Montgomery Ward
     law suit instigated
1892 Columbia Exposition

1900

1900s New wharves, piers Philadelphia
     and Seattle

1900s Lakefront park system begun
     in Chicago

1903 Olmsted brothers' park plan for
     Seattle includes waterfront park

1904 Baltimore Fire

1905 Warehouse sheds cut off
     New Orleans waterfront views

1906 San Francisco Fire

1907 Burnham's Chicago plan

1910

1911 Port of Seattle formed
1912 *Titanic* 822 feet

1917 World War I

1919 Chicago Lakefront Ordinance

1920

1922 Huston Navigation District formed

1880 Chelsea-Gansevoort
     pier development (to 1910)
1884 Corlears Hook Park

1890 U.S. sets new pierhead lines

1894 Riverside Park extension
     (to 1937)
1895 Public recreation piers

1898 Harlem River Speedway

1911 State order to remove
     surface tracks of RR
     below W. 59 St.

1916 Women's League for the
     Protection of Riverside
     Park (to 1937)

1921 Port of New York
     Authority established

1923 New York Central RR
     ordered to electrify

1927 Holland Tunnel

1929 Regional Plan of New York
     and Its Environs (to 1931)
1929 East River Drive (to 1940)

Table 1 (cont.)

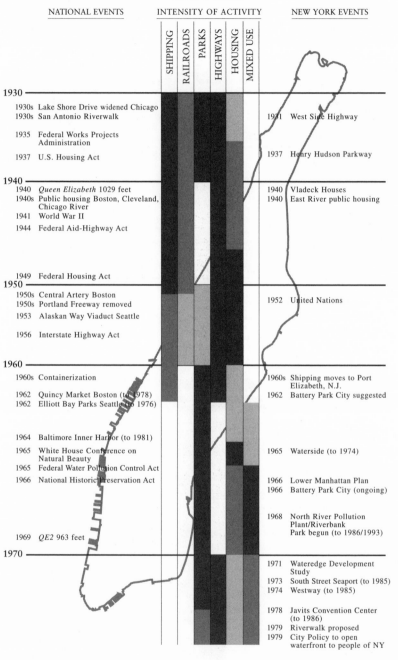

NATIONAL EVENTS	INTENSITY OF ACTIVITY						NEW YORK EVENTS
	SHIPPING	RAILROADS	PARKS	HIGHWAYS	HOUSING	MIXED USE	

1930

NATIONAL EVENTS	NEW YORK EVENTS
1930s Lake Shore Drive widened Chicago	
1930s San Antonio Riverwalk	1931 West Side Highway
1935 Federal Works Projects Administration	
1937 U.S. Housing Act	1937 Henry Hudson Parkway

1940

NATIONAL EVENTS	NEW YORK EVENTS
1940 *Queen Elizabeth* 1029 feet	1940 Vladeck Houses
1940s Public housing Boston, Cleveland, Chicago River	1940 East River public housing
1941 World War II	
1944 Federal Aid-Highway Act	
1949 Federal Housing Act	

1950

NATIONAL EVENTS	NEW YORK EVENTS
1950s Central Artery Boston	1952 United Nations
1950s Portland Freeway removed	
1953 Alaskan Way Viaduct Seattle	
1956 Interstate Highway Act	

1960

NATIONAL EVENTS	NEW YORK EVENTS
1960s Containerization	1960s Shipping moves to Port Elizabeth, N.J.
1962 Quincy Market Boston (to 1978)	1962 Battery Park City suggested
1962 Elliott Bay Parks Seattle (to 1976)	
1964 Baltimore Inner Harbor (to 1981)	
1965 White House Conference on Natural Beauty	1965 Waterside (to 1974)
1965 Federal Water Pollution Control Act	
1966 National Historic Preservation Act	1966 Lower Manhattan Plan
	1966 Battery Park City (ongoing)
	1968 North River Pollution Plant/Riverbank Park begun (to 1986/1993)
1969 *QE2* 963 feet	

1970

NATIONAL EVENTS	NEW YORK EVENTS
	1971 Wateredge Development Study
	1973 South Street Seaport (to 1985)
	1974 Westway (to 1985)
	1978 Javits Convention Center (to 1986)
	1979 Riverwalk proposed
	1979 City Policy to open waterfront to people of NY

Table 1 (cont.)

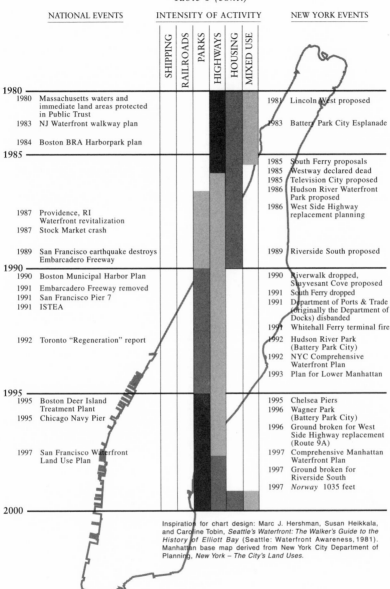

NATIONAL EVENTS	INTENSITY OF ACTIVITY						NEW YORK EVENTS
	SHIPPING	RAILROADS	PARKS	HIGHWAYS	HOUSING	MIXED USE	

1980

NATIONAL EVENTS	NEW YORK EVENTS
1980 Massachusetts waters and immediate land areas protected in Public Trust	1981 Lincoln West proposed
1983 NJ Waterfront walkway plan	1983 Battery Park City Esplanade
1984 Boston BRA Harborpark plan	

1985

	1985 South Ferry proposals
	1985 Westway declared dead
	1985 Television City proposed
	1986 Hudson River Waterfront Park proposed
	1986 West Side Highway replacement planning
1987 Providence, RI Waterfront revitalization	
1987 Stock Market crash	
1989 San Francisco earthquake destroys Embarcadero Freeway	1989 Riverside South proposed

1990

1990 Boston Municipal Harbor Plan	1990 Riverwalk dropped, Stuyvesant Cove proposed
1991 Embarcadero Freeway removed	1991 South Ferry dropped
1991 San Francisco Pier 7	1991 Department of Ports & Trade (originally the Department of Docks) disbanded
1991 ISTEA	1991 Whitehall Ferry terminal fire
1992 Toronto "Regeneration" report	1992 Hudson River Park (Battery Park City)
	1992 NYC Comprehensive Waterfront Plan
	1993 Plan for Lower Manhattan

1995

1995 Boston Deer Island Treatment Plant	1995 Chelsea Piers
1995 Chicago Navy Pier	1996 Wagner Park (Battery Park City)
	1996 Ground broken for West Side Highway replacement (Route 9A)
1997 San Francisco Waterfront Land Use Plan	1997 Comprehensive Manhattan Waterfront Plan
	1997 Ground broken for Riverside South
	1997 *Norway* 1035 feet

2000

Inspiration for chart design: Marc J. Hershman, Susan Heikkala, and Caroline Tobin, *Seattle's Waterfront: The Walker's Guide to the History of Elliott Bay* (Seattle: Waterfront Awareness, 1981). Manhattan base map derived from New York City Department of Planning, *New York – The City's Land Uses*.

Part One

... I have seen the hungry ocean gain
Advantage on the kingdom of the shore,
And the firm soil win of the watery main,
Increasing store with loss, and loss with store.

William Shakespeare, *Sonnet 64*.

CHAPTER I

Patterns

Setting the Shoreline, 1600–1950. The current revitalization of urban waterfronts is the latest of a long series of piecemeal changes that have affected New York and other North American port cities since their settlement. While occurring in a random fashion, these changes form a pattern that, in retrospect, is evident in the growth of waterfront cities from the Atlantic to the Pacific coast. Beginning in the seventeenth century in the East, in New York, Boston, Baltimore and Philadelphia; in the eighteenth century, in New Orleans, St. Louis and San Francisco; and in the nineteenth century, in Houston and Seattle, shipping determined the shape and use of the town shore. Initially, to handle such goods as lumber, grain or furs coming to town for export, and the cotton, raw sugar, or fine china imported for domestic use, the municipality built a small wooden pier supported on wood piles. Vessels known as lighters shuttled between an anchorage in the harbor and the dock.[1]

As trade increased and ships became larger, and space was

needed for storage of goods and fuel and for new industries such as shipbuilding and commercial fishing, piers grew both in number and size. By 1708 in Boston and neighboring Charlestown, seventy-eight wharves jutted out into the harbor. Most impressive, Long Wharf stretched eight hundred feet, allowing ships to unload without lighterage. Generally, private citizens built these docks for trading. Thomas Clark constructed what would become known as Lewis Wharf in Boston, and Johannes Beekman built Beekman's Wharf, now Fulton Street, in New York. Public inducements also helped. In Baltimore, for example, the state extended the city's pierhead lines in 1783 and again fourteen years later, enabling owners to add piles to the ends of their wharves to make them longer.

In the nineteenth century, from New York's East River to Seattle's Elliott Bay, the waterfront became the center of commercial activity and human contact. People came to these places to work in ship and lumber yards, on piers and in counting houses. They also came to purchase sails, food, and linen. They came to travel across river and ocean, to greet the newly arrived, and simply to watch the flurry of boats in the harbor.

During this lively period, sporadic modernization of piers continued. New technology was an important incentive. Eastern cities like Boston and Philadelphia sought to replace the trade lost to New York when the Erie Canal opened. They encouraged renovations to accommodate the newest vessels powered by steam instead of sail. Boston's Granite Wharf and accompanying warehouse, a pier made completely of stone, and the city's quarter-mile-long Central Wharf were responses.

Fires provided other opportunities for pier additions. In

Baltimore, Seattle, and St. Louis, conflagrations that caused the rebuilding of wooden cities with brick and stone also prompted the rebuilding of outdated wooden waterfront structures. The new piers that were thus built in Boston's South End in 1872 coincided with a demand for shipping space by the city's growing shoe and textile manufacturing businesses; the 1906 San Francisco fire motivated that city to rebuild its piers to suit the heavy, steel-hulled "leviathans" then coming into port.

By the mid-nineteenth century, streets, warehouses, and docks formed the edge of most port cities, and the railroad was adding a new dimension. By building tracks on existing streets behind the piers, the railroad boosted commerce. However, it began to remove the waterfront from public view. The Baltimore and Ohio Railroad led the way in 1828, laying tracks from the port on the Patapsco River to areas south and west of Baltimore. In the 1860s railroads from the north of Boston acquired access to piers along Atlantic Avenue. A decade later, local railroad lines serving Philadelphia ran to piers on the Delaware and Schuylkill rivers, and transcontinental lines terminated at the ports of San Francisco and Oakland. The railroad made its presence known in similar ways in other cities. Eight tracks paralleled the Mississippi in New Orleans by 1880, and in some places up to sixteen tracks formed a barricade along Lake Ontario in Toronto that reduced both visual and pedestrian access to the water.

The railroad also set new outer limits to many shorelines. In Chicago the Illinois Central was given a two-mile-long, three-hundred-foot-wide strip of land between Michigan Avenue and the lake in 1856. In return for receiving water rights to build a trestle over the lake leading to its new station, the Central agreed to construct a breakwater (which the city could not afford to do) to halt erosion of the shore. Double tracks soon lined the lakefront from center city south to 51st Street.[2]

In addition to expanding the limits of the land, piers and railroad trestles also became foundations for new real estate parcels. When steep hills prevented nineteenth-century expansion inward in Seattle, the land was simply extended at wharf-ends. Ships dumped rock ballast into Elliott Bay, widening and lengthening adjacent landings while at the same time making room in their holds for the lumber and coal destined for San Francisco and markets around the world. The Seattle and Walla Walla Railroad trestle eventually provided a massive retaining structure behind which mud flats and later portions of Elliott Bay could be filled in with earth from building excavations and street gradings, wood waste from lumber mills, household refuse, and dredge material.[3]

Twentieth-century technical improvements added still another stratum and a new structure to urban waterfronts—the elevated highway. The assembly line, balloon tire, and better roadbeds led to the widespread use of cars and trucks after World War I. Traffic congestion, formerly most critical on downtown waterfronts, escalated in the city's interior business and residential streets. The shoreline provided cheap land on which to find relief with minimal dislocation. As a result, Chicago's Lake Shore Drive, opened for horse drawn carriages in the 1870s, was widened and extended at grade from 13th to 56th Street along Lake Michigan sixty years later. In Seattle and Toronto the railroad had a firm foothold on the outer street. On Boston's Atlantic Avenue an elevated railroad cast dismal shadows. There was no place to build highways but up. From the Atlantic to the Pacific, elevated expressways completed the physical, visual, and psycholog-

ical separation of cities and their residents from the waterfront, which had begun a century before.

Administering the Shore, 1700–1950. As demands for the use
of the waterfront increased, jurisdictional responsibility also
shifted. In American cities, land under the water adjacent to
the shore originally belonged to the state. In growing sea-
ports, riparian rights were often transferred to the city, which
either leased or sold underwater land to private individuals
for development. By the mid-nineteenth century, landowners
and shippers alike were begging the state to reassert its authority to help with aging and outmoded waterfront structures. In San Francisco, where precipitous growth caused by
the discovery of gold required immediate attention, the state
answered by constructing a two-mile-long sea wall from
which new private piers, warehouses and railway switchyards
soon extended.[4]

Not until 1879 did the federal government have a formal
role in local waterfront development. In that year the Rivers
and Harbors Act authorized the United States Army Corps
of Engineers to regulate all activities in the nation's navigable
waters.[5] This legislation gave the army sole power to set new
bulkhead and pierhead lines. In addition, it could make certain
harbor improvements if federal funds were so designated. A
decade after receiving its powers, the corps defined crucial
new pierhead lines in New York. This agency helped Boston
to cope once again with technological change. Hope for that
seaport, which had never recovered from the opening of the
Erie Canal, was revived when, between 1892 and 1905, the
army dredged entry channels to accommodate wider ships
with deeper drafts.[6]

At the turn of the century, municipal agencies with condemnation powers assumed responsibility for port expansion in many cities. By subsuming many of the roles formerly delegated to private developers, these departments had significant effect on the physical profiles of several downtown waterfronts. In 1905 the New Orleans Dock Board, for example, built new steel transit sheds to store merchandise arriving on the nineteenth-century wharves. While modernizing shipping facilities, these warehouses, packed end-to-end, effectively cut off any remaining visual links between the city and the Mississippi River. In Philadelphia, the impact of the newly created Department of Wharves, Docks and Ferries was different. This agency doubled the city's shipping tonnage by building longer and more modern piers on the city's central and south Delaware riverfronts. Ironically, the latter, more desirable, one-thousand-foot-long installations drew marine commerce away from the new, but shorter, piers downtown, leading to the eventual abandonment of the central facilities.[7]

Port commissions and authorities were also formed in the early decades of the twentieth century to solve some of the problems caused by jurisdictional ambiguities and the confluence of piers, streets, and railroad tracks on downtown waterfronts. An independent navigation district was created in Houston in 1922, when neither the city nor the overlapping county had sufficient power to carry out harbor improvements. The first issue confronting the Port of New York Authority, a bistate body (New Jersey and New York, known as the Port Authority), was the chaos caused by a single railroad line that occupied major waterfront thoroughfares and by other carriers which were dependent solely on lighterage services to move goods in and out of the island of Manhattan. The new Port of Seattle was faced by another unique problem: how to remove the congestion and the switching confusion caused by four transcontinental railroad companies vying for space on that city's downtown bayfront.

The federal government began to expand its programs for financing urban waterfront developments during the depression, and at the same time influenced changes in the types of activities to be created there. The shoreline with its slums, cheap land, and open views provided a perfect setting to spend Works Progress Administration funds to provide jobs and housing and to clear blighted areas. Extension and widening of Chicago's Lake Shore Drive and groundbreaking for New York's East River Drive provided work for the unemployed of those cities. In addition, Cleveland's Lakewood project on Lake Erie, Boston's Maverick Street development a block from Boston Harbor, and New York's Vladeck Houses, separated from the East River only by a highway, provided subsidized dwellings to replace blighted waterfront neighborhoods. After World War II, federal funding programs such as the Federal Housing Act and the Federal Aid to Highways acts had a further impact on the urban shore. The high-rise public housing projects and the expressways that rose on the rivers and bays of such cities as Buffalo and Boston were the result.

Chicago, 1800–1940. Most waterfront cities in North America show similar development patterns. Chicago, however, is an exception. A continuous park, bordered in places by luxury housing—not wharves, storage sheds, or freight tracks—ultimately settled on the edge of Lake Michigan. A plot of land four blocks square, subject to flooding and therefore undesirable for prospective homes or industry, set the pattern for recreational development of Chicago's lakefront. In 1836 commissioners of the Illinois and Michigan Canal Company were charged with selling unsettled areas of the tiny municipality of Chicago to pay for construction of a canal connecting Lake Michigan to the Mississippi River and the Gulf of Mexico. They mapped twenty useless acres between Michigan Avenue and the lake, Madison to 12th Street, as a public ground. According to the commissioners, this was to be a "common to remain forever open, clear, and free of buildings or other obstruction whatsoever."[8]

The transition from swampland to parkland was not smooth. As the city grew, the site near the mouth of the Chicago River and adjacent to the downtown was in demand for other municipal uses. In the 1840s construction of the Illinois Central Railroad trestle actually widened the area, adding to the underwater land in Lake Michigan. In the ensuing decades this was filled in further with ashes, garbage, and the debris from the Great Fire of 1871. It was not an attractive park. A motley array of railroad sheds, stables, a firehouse, and squatters' shacks soon dotted the fill; a late nineteenth-century civic improvement scheme aimed to add a city hall, post office, police station, power plant, and new stables for the city's garbage carts.

The efforts of one man finally did establish a park on Chicago's downtown lakefront. One day in 1890, as the story goes, Aaron Montgomery Ward, founder of the nation's first mail-order house, looked down from his skyscraper window on the mess that was the common and filed suit to clear the lakefront.[9] Eleven years and several court suits later, Grant Park was opened, with the addition of fifty new acres on fill, east of the Illinois Central tracks. Chicago had its first lakefront recreation space and, in addition, a presentable yard for elegant new homes and apartment houses.

By the turn of the century, isolated patches of green were

not unknown on otherwise industrial waterfronts. In the case of Chicago, however, Grant Park became part of a solid line of lakefront parks. Two events in the 1860s helped to move heavy industry, dumps, and coal landings away from Lake Michigan to free it for subsequent recreational spaces. The stockyards were consolidated on the southwest side of the city, where they were amply supplied with railway connections and no longer needed the lake for transportation. Then, in 1869, the federal government allocated money to improve a harbor on the Calumet River to the south of the city. This spared the lakefront from a proliferation of bulk cargo facilities and rail and truck terminals. It also caused a gradual decline in shipping in the Chicago River harbor on the northern lakefront, a few blocks above Grant Park.

As was the nineteenth-century pattern elsewhere, the Illinois Central Railroad tracks effectively separated Chicago from its shorefront, yet unlike the experience elsewhere, it also contributed to park development. The company's passenger station on the southern end of the city became a reason for the Parks Commission to lay out a second waterfront recreation space, Jackson Park. In 1919 the Lake Front Ordinance, an agreement between the city and the railroad, laid the legal foundation for connecting Jackson Park to Grant Park. While the railroad promised to switch from steam to electricity and eventually to remove noise and smoke from the waterfront, the Parks Commission was given the authority to create new beaches and parkland to the east of the railroad tracks on eight miles of the shore.[10]

With a waterfront ripe for commercial or residential development, the fledgling city planning movement added its contribution to the greening of Chicago's shore. The Columbian Exposition and World's Fair of 1893 provided the funds and impetus to complete Jackson Park, which had been de-

19

signed by Frederick Law Olmsted two decades earlier. The fair also laid the philosophical groundwork for Daniel Burnham's *Plan of Chicago,* made public in 1909, which highlighted and reinforced the natural beauty of the lakefront. The lagoons and harbors formed by offshore islands envisioned by Burnham were never built. However, by 1940 other legislation and an infusion of federal funds to combat the depression had created a luxurious residential neighborhood and a continuous public playground stretching from the northern to the southern end of the city.

By the beginning of World War II, with few exceptions, dozens of North American waterfront cities were bound not by their waterways but by the very structures that enabled them to function economically and socially. Layer by layer, facilities had been built to move people and cargo and to provide a place to live. From the Mississippi to the Great Lakes, New York Harbor to San Francisco Bay, piers and pier sheds, railroad tracks and sidings, river level and elevated highways, and apartment buildings for rich and for poor formed multilevel barriers to the eye and to the pedestrian. To understand in even greater detail just how this happened, step back in time and observe the development of Manhattan's shoreline.

NOTES

1. The following publications provided useful information on North American waterfront cities: George Brambilla and Gianni Longo, *Learning from Baltimore* (Washington, D.C.: Institute for Environmental Action, 1979); Brambilla et al., *Learning from Seattle* (Institute for Environmental Action, 1980); Jane S. Brooks, Richard O. Brumback, Jr., and Susan Drake, "The Resurgence of Urban Waterfronts: Redevelopment Along the Mississippi River's Edge in New Orleans," paper presented at the 1985 Annual Meeting of the Urban Affairs

Association, Norfolk, Virginia, April 1985; City of Boston, *Harbor-park: A Framework for Discussion* (Boston: Boston Redevelopment Authority, October 1984); City of Philadelphia, Philadelphia City Planning Commission, *Central Riverfront District Plan,* n.d.; Committee on Urban Waterfront Lands, *Urban Waterfront Lands* (Washington, D.C.: National Academy of Sciences, 1980); Ann Breen Cowey, Robert Kaye, Richard O'Conner, and Richard Rigby, *Improving Your Waterfront: A Practical Guide* (U.S. Department of Commerce, National Oceanic and Atmospheric Administration, 1980); Virginia Farrell, *Development and Regulation of the Urban Waterfront: Boston, San Francisco and Seattle* (Princeton: The Center for Energy and Environmental Studies, Princeton University, 1980); Maurice Freedman, "The Special Problems of Waterfront Development," paper presented at Waterfront Development Conference, New York University Real Estate Institute, New York, March 1986; Freedman, "Waterfront Revitalization in a New Physical Era, *Real Estate Review* 11 (Winter 1982):77–82; Arlene Gemmil, *Ontario Place: The Origins and Planning of an Urban Waterfront Park* (Department of Geography, York University, Toronto, Canada, April 1981); Clare Gunn, David J. Reed, and Robert E. Couch, *Cultural Benefits for Metropolitan River Recreation—San Antonio Prototypes* (Technical Report No. 43, Texas Water Resources Institute, Texas A&M University, June 1972); Anna Marie Hager, "A Salute to the Port of Los Angeles: From Mud Flats to Modern Day Miracle," *California Historical Society Quarterly* 49 (December 1970):329–35; Andy Leon Harney, ed., *Reviving the Urban Waterfront* (Washington, D.C.: Partners for Livable Places, National Endowment for the Arts, and Office of Coastal Zone Management, n.d.); Marc J. Hershman, Susan Heikkala, and Caroline Tobin, *Seattle's Waterfront: The Walker's Guide to the History of Elliott Bay* (Seattle: Waterfront Awareness, 1981); Marc J. Hershman and Robert Warren, eds., "Theme Issue: Urban Coastal Management," *Coastal Zone Management Journal* 6 (1979); Heritage Conservation and Recreation Service, *Urban Waterfront Revitalization: The Role of Recreation and Heritage,* 2 vols. (Washington, D.C.: Department of the Interior, n.d.); Mayor's Riverfront Committee, *The Saint Louis Central Riverfront: An Analysis of Challenges and Opportunities* (September 1984); Sherry Olsen, *Baltimore: The Building of an American City* (Baltimore: Johns Hopkins University Press, 1980); Marilyn McAdams Sibley, *The Port of Houston: A History* (Austin: University of Texas Press, 1968); Sam Bass Warner, *The Urban Wilderness: A History of the American City* (New York: Harper and Row, 1972); Walter Muir Whitehill, *Boston: A Topographical History* (Cambridge: Harvard University Press, 1959); Workers of the Writers' Program of the Work

Projects Administration in the State of Massachusetts, *Boston Looks Seaward: The Story of the Port 1630–1940* (Boston: Bruce Humphries, 1941); and Douglas M. Wrenn, *Urban Waterfront Development* (Washington, D.C.: Urban Land Institute, 1983). See also "Waterfront Plans, Reports, and Studies—General" in Selected Bibliography.

2. Information on Chicago comes from the following sources: Art Institute of Chicago, *The Plan of Chicago, 1909–1979* (Chicago, 1979); City of Chicago, Chicago City Planning Commission, *The Lakefront Plan of Chicago* (Chicago, December 1972); Carl W. Condit, *Chicago 1910–1929: Building, Planning, and Urban Technology* (Chicago: University of Chicago Press, 1973); Carl W. Condit, *Chicago 1930–70: Building Planning, and Urban Technology* (Chicago: University of Chicago Press, 1974); Constance McLaughlin Green, *The Rise of Urban America* (New York: Harper and Row, 1965); Harold M. Mayer and Richard C. Wade, *Chicago: Growth of a Metropolis* (Chicago: University of Chicago Press, 1969); and Lois Wille, *Forever Open Clear and Free: The Historic Struggle for Chicago's Lakefront* (Chicago: Henry Regenery, 1972); Condit, *Chicago 1910–1929*, 22–23.

3. Hershman, et al., *Seattle's Waterfront*, 22–23.

4. Wrenn, *Urban Waterfront Development*, 102.

5. Ibid., 18.

6. Writers' Program, *Boston Looks Seaward*, 150, 179.

7. City of Philadelphia, *Central Riverfront District Plan*, 7.

8. Wille, *Forever Open Clear and Free*, 23.

9. Ibid., 73–74.

10. Meyer and Wade, *Chicago*, 100–102, 294.

CHAPTER 2

Foundations for the Wall

The island of Manhattan found by the early explorers was what remained above water of a high plateau extending 150 miles inland from its current position in the Atlantic Ocean. In four successive waves, ice masses shaved the mountains and incised river and harbor beds. When in time the waters rose, there remained a low-lying island punctuated by hills and surrounded by a river, two estuaries, and a large, protected, deep-water bay.[1]

The Shores of Nieu-Nederlandt and New York, 1624–1690. Because of landfill, Manhattan south of City Hall is today 33 percent larger than when Peter Minuit bargained with the Indians for its control. Geographically, Manhattan is an island and there is a vacuum at the edge into which, until recently, it was assumed expansion could occur. Ballast was dumped and ships sunk, hills leveled, building sites and roadways excavated, wastes, ashes and sweepings collected, and all were deposited at the water's edge. When space was needed for services, work places, homes, or recreation, it was always possible to create more land.[2]

Manhattan's boundaries, as they have been reconstructed in the southern portion, were at the original high-water mark: Pearl Street on the narrower, sheltered East River and Greenwich Street on the broader, less turbulent but ice-prone Hudson. (Map 1) Tracts of land belonging to the city's "first" families, among them Stuyvesants, Lispenards, and Van Curlers, stretched to the edges of the island. In places the shore was flat and marshy, in others it was formed by high, rocky cliffs. Elsewhere, gentle, flowering meadows sloped to the water's edge.

The island's western shoreline, with at times a mile of water separating it from mainland New Jersey, followed the line of Washington Street from Reade to West 13th Street (except for a bulge out to West Street between Clarkson and West 11th streets). From West 14th Street northward to West 30th Street the Hudson lapped against the island at Tenth Avenue and then meandered between Eleventh and Twelfth avenues to West 60th Street. From there, to the northern terminus of the island, what became Twelfth Avenue defined the outer limit of the city.[3]

On the East Side the swift East River currents swept by the western edge of Cherry Street, from Dover to Jefferson streets, and split the island of Manhattan from the larger island of which Brooklyn is a part. The bulge at Corlears Hook was pierced by a small inlet between Delancey and Houston streets. The area currently called Stuyvesant Cove (between East 12th Street and Bellevue Hospital at East 25th Street) was a natural cove that curved to the west in some places as far as First Avenue. The adjoining land was wet, permeated by a marsh and a series of rivulets feeding into

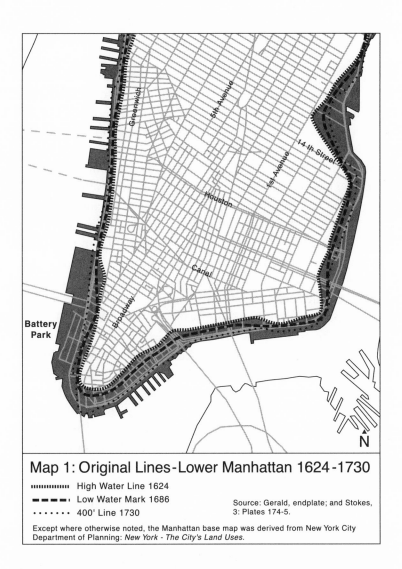

Map 1: Original Lines - Lower Manhattan 1624-1730

||||||||||| High Water Line 1624

▬ ▬ ▬ ▪ Low Water Mark 1686

• • • • • • • 400' Line 1730

Source: Gerald, endplate; and Stokes,
3: Plates 174-5.

Except where otherwise noted, the Manhattan base map was derived from New York City
Department of Planning: *New York - The City's Land Uses.*

the East River. North of the farm from which Bellevue derived its name to the rocks of Hellgate (opposite East 89th
Street) the river created a boundary between Manhattan and
Queens. Here, on the Manhattan side, many of the original
limits of the land remain, the banks merely straightened by
upstream sediments and by the hand of man. Then at East

90th Street swift currents, pushed by the higher tidal range in Long Island Sound, excised still another cove. This, one of the few remaining recesses in an otherwise straight edge, was a marshy indentation extending almost to First Avenue.

The banks of Manhattan's third and narrowest body of water, the Harlem River, have been moved but only minimally expanded. (Map 2) To keep this river (actually an estuary) intact for shipping, its course was changed and its depth increased. In the seventeenth century, the river began at Hellgate and, while separating Manhattan from the Bronx (Westchester) mainland, meandered along its present course in a northwesterly direction to West 155th Street. Here, in the nineteenth century, a small group of islands were merged into a speedway, today a highway and park. Above this, the land on the edge of the river was fairly precipitous, pierced by a single inlet, between Dyckman Street (Inwood Street) and West 202nd Street, into which flowed Sherman Creek (Half Kull). A park and housing project now fill this cove. From Sherman Creek, the Harlem River continued northward another mile to West 222nd Street, where a small stream carved out a six-block island at the tip of the island of Manhattan known as Marble Hill.

At West 228th Street the Harlem River curved westward to the top of Manhattan (and Marble Hill) where it joined with Spuyten Duyvil Creek. The course of this tributary, altered radically in the 1890s when the Harlem Ship Canal was constructed, originally connected the Harlem and Hudson rivers. Before its diversion, the creek flowed southward from its juncture with the Harlem River to the bottom of the marshy Spuyten Duyvil peninsula opposite West 215th Street. This projection of the Bronx (Westchester) is now Inwood Hill Park. From here the creek curved northwesterly to its mouth at West 220th Street and the Hudson River.[4]

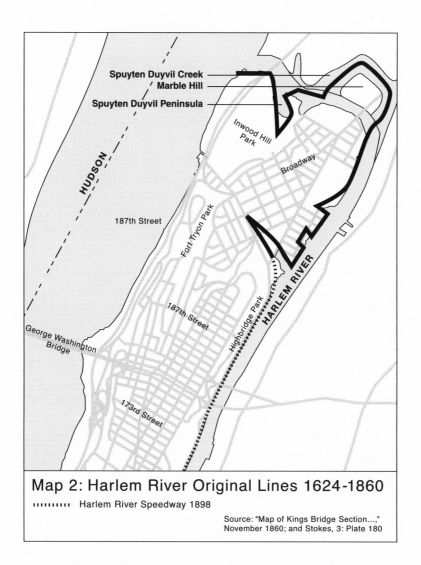

Map 2: Harlem River Original Lines 1624-1860

░░░░░░░ Harlem River Speedway 1898

Source: "Map of Kings Bridge Section...,"
November 1860; and Stokes, 3: Plate 180

The first changes in Manhattan's geographical profile occurred in the seventeenth century, primarily along the lower
East River. In 1623 the Dutch West India Company, formed
by merchants to capture the trade in the Americas, received
a grant for all of the land on the island of Manhattan (Nieu-

Nederlandt). The high-water mark (or high tide) set the boundary of the community. After keeping several plots for its use and a fortification, the company parceled out lots to the early settlers to develop privately for homes and gardens. Because the owners were typically merchants and fur traders who required access to the harbor, the majority of these grants were along both rivers at the southern tip of Manhattan.[5]

In 1647 public development, typical of eastern settlements like Boston and Baltimore, began in lower Manhattan. A simple dock, probably made of wood pilings topped by a platform of wood planks, was constructed into the East River at Schreyer's Hook (Pearl and Broad streets). Thereafter, incoming Dutch and foreign sailing vessels—small, one-masted sloops and two-masted, square-rigged brigs and ketches—anchored in the river and transferred their cargo and passengers by small boat to this public wharf. Twelve years later, in 1659, a second pier (the Bridge or Weighouse Pier) was constructed a block to the west at the juncture of Whitehall and Pearl streets (today Moore Street).[6]

Although the West India Company made dock improvements beyond the high-water mark, the city did not actually own this property until after the British occupation in 1664. The city was renamed New York, and for two decades the land under the rivers belonged to the Crown. Then, on 22 April 1686, Lt. Governor Thomas Dongen transferred ownership of unencumbered lands to the City of New York. "All waste, vacant, unpatented, unappropriated" lands and water bodies within the island to the low-water mark (or low tide) were granted to the corporation, as well as title to public spaces such as the wharves, docks and "every street, lane, highway and alley." Particularly important was the extension of the city limits from the high- to the low-water mark and the broadened municipal powers over these lands. As the corporation could now "fill, make up, lay out, use and build on" lands under water, the city could expand, in places up to two hundred feet, or one block.[7] (Table 2)

With a superb, natural, deep-water harbor and landings in the East River protected from wind and ice floes, shipping became the mainstay of the city's seventeenth-century economy; and as trade increased in the small seaport it was profitable to develop the edge. Thus, the city quickly disposed of the newly acquired space between high and low water. These lots were sold with the proviso that the owner must build the street and wharf along the water end, for there was no public agency to provide these structures. On the East Side, a year after the Dongen Charter (1687), the five water-edge blocks between Whitehall and Old Slip were sold. The area from high water at Pearl Street to the low-water line at Water Street was filled, and a new wharf was constructed at the edge to retain the land. Dutch piers at the foot of Broad and Whitehall streets were by then thirty-nine and twenty-seven years old and at the far end of the useful life of a wooden waterfront structure. These probably provided either a retaining wall or waste material for this expansion.[8]

More extensive public works also affected the shape of the East River waterfront. Anxious to increase their colonial exports, the British constructed the Great Dock in 1675. Consisting of curved stone extensions of Coenties Slip and Whitehall Street, which together formed a large, protected, wet basin two blocks outshore of the original high-water line, the Great Dock also enclosed two new waterfront blocks between Pearl and Water streets.[9]

Increasing pressure to house city residents, who had grown in number to 4,476 by 1700, created a need to find a place to deposit excavated earth. The canal at Broad Street (De

Table 2. Manhattan Landfill: 1686–1984

Date Begun or Proposed	Legal Boundaries			Actual Made Land	
	Legislation	Boundary	Acres	Location	Acres
1624	Minuit's purchase	Manhattan Island to high-water line	1200		
1686	Dongen Charter	High– to low-water line around island	NA	East River Whitehall to Corlears Hook	95
				Hudson River, Battery to West 32nd Street	226
1730	Montgomerie Charter	400 feet east of Water Street, Whitehall to Corlears Hook; and 400 feet west of Washington Street, Battery to Charlton Street	127 82.5		209.5
1798	Outer Streets and Wharves Act	South and West streets beyond 400–foot line			198.25
1803, 1826	Ordinances	400 feet east of low-water line around island	NA		NA

Table 2 (cont.)

Date Begun or Proposed	Legal Boundaries			Actual Made Land	
	Legislation	Boundary	Acres	Location	Acres
1821	Ch. 172 L. 1821	600 feet into the Bay and East and Hudson rivers		Battery	20 +
1871	Dock Department Plans	1870 pier and bulkhead lines		Bulkhead and pier expansion around island	300 +
1894	Ch. 152 L. 1894	Twelfth Avenue to 500 feet west of high-water line at West 72nd Street		Riverside Park, West 72nd to 129th Street	132
1898	Ch. 102 L. 1893			Harlem River Speedway, East 155th to Dyckman Street	NA
1937–39				East River Drive, Montgomery to East 93rd Street and associated parkland	40 52
1961–73	Ch. 994 L. 1970	Bulkhead to pierhead line		Waterside, East 25th to 30th Street	6.1 (deck)

Table 1 (cont.)

Date Begun or Proposed	Legal Boundaries			Actual Made Land	
	Legislation	*Boundary*	*Acres*	*Location*	*Acres*
1961– 76	Ch. 994 L. 1970	Bulkhead to pierhead line		Battery Park City, Battery to Harrison Street	91
1962– 78				North River Pollution Control Plant, West 137th to 145th Street	30 (deck)
*Approximate total landfill 1980 from published figures					1399.85
Office of Manhattan Borough President *Estimate of Landfill 1640 to 1940*					
		East River		600	
		Hudson and Harlem rivers		1500	2100
Landfill and decking 1940–1980					127.1
*Estimated total landfill 1980					2227.1

Sources: Board of Assistant Alderman, Doc. 9, 9 October 1848, frontispiece; Borough President of Manhattan, *Annual Report,* 1 July 1941, 18; Robert Caro, *The Power Broker,* 557; Daniel Ewen, *Survey of the Battery* (McSpedon & Baker, September 1848); Stanley Isaacs Papers, File Box 5 "East River Drive 1940," New York Public Library; Richard Baiter, Office of Lower Manhattan Development, *Lower Manhattan Waterfront,* 9; New York City Planning Commission, Comprehensive Planning Workshop, *The New York City Waterfront,* June 1974, 91; New York State Office of Parks, Recreation and Historic Preservation; Stokes, *Iconography,* 5:1812–13; International Congress of Navigation, Trip of Inspection around New York Harbor, 4 June 1912; and *Valentine, Manual for 1856,* 598.

*Accurate landfill figures are difficult to compile because in so many instances acreage was never recorded. For example, the only acreage count to be found after 1862 (Isaacs, 1940) listed the East and Hudson river landfills in two lump sums. How much of this belonged to the Harlem River or included deposits of earth in swampland, was not calculated.

Two totals are therefore listed. The first shows landfill primarily at the lower half of Manhattan (with the exception of Riverside Park and the Pollution Control Plant). The second, "Estimated Total Landfill 1980," is a close approximation to the contemporary shoreline, probably including filled streams and swamps.

Heere Gracht), which in its Dutch guise had provided the first safe harbor for commerce, was filled with the diggings of building and street foundations. On the new-made land, Broad Street was elongated one block into the East River.[10]

In retrospect, the Dongen Charter offered New York a unique public planning opportunity. With ownership and governance powers over the waterside the city had the potential to decide what and where to build. However, the concept of public planning was not indigenous to this society; as New York grew, it continued the practice begun by the Dutch, and both ownership and development decisions were transferred to private individuals.

Hudson River Lethargy, 1700–1776. Manhattan pushed out into the adjacent rivers in spurts. For decades only minor changes occurred and then, in a single year when fill was suddenly available, concaves became convex and the inhabited shoreline elongated. On the East Side, when the hills flanking Maiden Lane were leveled to allow the city to expand northward, owners of water lots found cheap material to augment their property. Thus the edge between Hanover Square (Old Slip) and Catherine Street was extended eastward to Water Street.[11]

In enlarging the capabilities of their land, entrepreneurs, merchants as well as craftsmen, added both to the infrastructure of the city and to the variety of profitable activities taking place along the East River. North of Wall Street, shipwright Joseph Latham, joiner John Ellison, and others whose names read like the Social Register—de Peyster, Van Cortlandt, Rutgers, and Livingston—continued Water Street as the outer replacement for Pearl Street and added private docks. In the first two decades of the eighteenth century,

33

houses and stores sat on East River banks protected by stone bulkheads, and on wooden stilts projecting over the water. By 1717 four shipyards stood on the unimproved beaches.[12]

Private expansion to the low-water line now began along the Hudson River, however on a much smaller scale. The earliest grants were made between 1699 and 1701 to "P. J. Meiser and others," who acquired approximately three blocks between Cedar and Cortlandt streets from the shore to the low-water mark. (Map 3) On this land, three blocks above Wall Street and north of the waterfront area being developed on the East River, individuals with disparate interests broadened their local residential and commercial holdings. John Rodman, dock owner, John Hutchins, proprietor of the Coffee House Hotel, and William Huddleston, founder of the Trinity School, added some fill and rudimentary docks to the original line of the Hudson River shore.[13]

In 1705, the possibilities for expansion two blocks to the north, between Fulton and Warren streets, were expedited. The Governor, Lord Edward H. Cornbury, granted the Queen's Farm, which ran east to west between Broadway and the Hudson River, to the Trinity Church. (Map 3) Within three decades, this portion of the shoreline would be available for development, and the Meiser area would be filled in with private docks projecting into the river at Albany, Cedar (Little Green) and Cortlandt streets.[14]

Curiously, this first West Side landfill skipped the entire southern section of Hudson River waterfront from the Battery to Rector Street (the lower end of Battery Park City). This area, already well developed with homes and businesses, remained untouched for nearly a century. (Map 3) This strange circumstance could be explained by the dominance of the East River as the established business and shipping

34

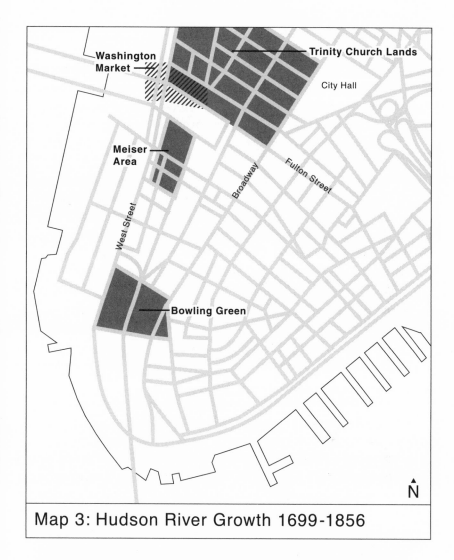

Map 3: Hudson River Growth 1699-1856

center; it already provided more than enough room and shelter for the demands of the eighteenth-century seaport.[15]

Another deterrent to the expansion of the Lower West Side of Manhattan into the Hudson was the depth of the water, which in places touched hard rock at forty feet. Such deep-water landings were not required by the small, wind-driven boats then in use. Furthermore, any pier or bulkhead construction would entail complicated and expensive engineering techniques—pile-driving, extensive fill, and durable retaining walls. An undertaking of this sort would have to await power-driven equipment and a ready supply of cement before it was financially reasonable.

This area was also slow to develop because the waterfront was privately owned by a few wealthy individuals, who enjoyed their largely residential neighborhood and who could influence the shape of the waterfront, but only when they were ready. Beginning in 1734, properties along the river side of Broadway were purchased and "handsome residences" erected.[16] Number 1 Broadway became first the house of Abraham de Peyster, a former Mayor of the City of New York, and later the home of the British Naval Captain Archibald Kennedy. Riverside views and nearby Bowling Green and the Lutheran Church added to the residential desirability of the neighborhood.

These owners too had no pressing economic need and could afford to wait until market conditions ripened and the formation of new blocks and lots or wharves and piers was profitable. In 1765 attorney Augustus Van Cortlandt acquired the water rights behind his family's house at Number 9 Broadway, and a large part of the soil and water beyond Greenwich Street, between Morris and Rector streets, was ceded to the heirs of Sir Peter Warren and to his brother-in-law Oliver Delancey. Ownership privileges, increased under the Montgomerie Charter of 1730, now extended four hundred feet, or two blocks, beyond the low-water mark. When issuing these grants, the city included the proviso that three streets be built parallel to the river (Greenwich, Washington and West streets). Yet nothing was done; these wealthy owners simply sat on their holdings.[17]

Although the Montgomerie Charter increased the opportunities for waterfront development, the southernmost Hudson River shoreline continued to remain in its natural state. By the late 1700s, when the city applied to the legislature to allow the widening of West Street, it claimed that it needed a means to compel owners to build streets when they were granted water rights. Only Greenwich Street was complete. West Street had not proceeded beyond surveyors' maps and the newest outer street, Washington Street, ended a few feet above Morris (Beaver) Street. Below this, to Battery Place, the river rose to a few yards west of the Kennedy property on Greenwich Street.[18]

Eighteenth-century expansion into the Hudson River continued to the north instead, where there were few established homes, where land was cheapest, and where there was room for expansion. Here, as the Trinity Church lands were sold (beginning in 1773 from Vesey to Chambers Street), was space not only for piers but for the ancillary activities, warehouses, and shops that had shaped the prosperous East River banks. At Cortlandt Street there was even a small ferry that since 1661 had carried provisions and people between Hoboken (Bergen) and Manhattan.

Revolution and Retrenchment, 1776–1807. New York's policy of liberal waterfront grants to private individuals, with covenants that required street and wharf improvements, failed to create a shoreline that kept pace with the growing commercial needs of the eighteenth century. The number of vessels frequenting the port grew exponentially while the pier space lagged far behind. In 1699 city residents owned sixty ships and a variety of sloops and boats. By 1720 Boston was the principal mart of North America, but New York had caught up to its other shipping rival, Philadelphia. Forty years later vessels belonging to New Yorkers (which now included the two-masted schooners in vogue for the coastal trade) had increased sevenfold in number to 447. In a decade, with only two public wharves and a half-dozen private piers to serve them this number grew to 709.[19]

The Revolution merely postponed a solution to the dock shortage. Shipping was curtailed by British taxation, resultant boycotts, and finally wartime occupation of the harbor. Waterfront construction stopped and piers deteriorated. No new land was made, for real estate was an unprofitable venture, and such economic activity as there was in the wartime city did not require additional fill. The population of Manhattan dropped from 20,000 in 1776 to 10,000 at the end of the war. Two huge fires during the conflict demolished one quarter of the buildings, while many others were destroyed by pillage or for use as winter fuel.[20]

Recovery was swift in the plucky city. In less than a decade, the population returned to its prewar high. By 1790 new trade routes had been opened between New York and China. Imports rose from one to two million dollars in three years, tonnage doubled, and the population increased by a third to 33,131.[21]

With revived prospects of income from waterfront properties, both the city and private individuals were buying, filling, building, and repairing. Results could be seen at the East River as far north as Corlears Hook. By 1800 parts of this shore were filled out even beyond the four-hundred-foot limit set in the Montgomerie Charter, or to South and Front streets. On the Hudson River between Vesey and Fulton streets, the city purchased water grants it had so readily sold earlier and so established itself as an active waterfront realtor.

The piers were lengthened and the slip between them was filled in to expand a profitable market.[22]

The 1789 Outer Streets and Wharves Act codified municipal displeasure with private control of the water's edge. As a result of the individualistic method of building, together with the postrevolutionary boom in Manhattan, the existing bulkhead had become extremely irregular. According to J. W. Gerard, Jr., an attorney who in the late nineteenth century represented many riparian owners in the sale of their water lots back to the city, the late-eighteenth-century shoreline was "without settled plan, and followed the sinuosities of the river; the grants extending to unequal distances into both rivers."[23]

The act helped the city to make corrections. To remove the island's ragged edges, to guide the placement of landfill, and to provide straight pathways for travel and commerce, South and West streets were created. Rather than follow the natural curves, surveyors marked several points in the East and Hudson rivers and drew straight lines between them. On the West Side the resulting thoroughfare was within the four-hundred-foot mark. On the East River expansion was now sanctioned to 420 feet beyond the original low-water line at Roosevelt Street and up to 590 feet at Rutgers Slip.[24]

To avoid large, unfinished gaps in the outer streets and wharves, the act imposed more stringent rules on private owners. As in the past, waterfront lot owners who already abutted the newly mapped outer street were required to create it by filling and by building wharves. Holders of lots that had not been improved to the new exterior street lines were simply required to fill and level the spaces between their land and the edge of the new wharf.

As a final resort, even the city was allowed to make corrections. To ensure that these outer streets were constructed, the corporation was given the power to fill the "intermediate" spaces if proprietors failed to do so, charging the expense to the owners. Thus, another mechanism was put in place to grant the city an increasing role at the water's edge.[25]

The Randall Plan or Commissioners' Map, ordered by the State Legislature in 1807 and published in 1811, placed additional public controls on the city's shores. This official map of the city left undisturbed the then-developed portions of Manhattan, which ran from Corlears Hook on the East River to Jay Street on the Hudson. North of this, it imposed a tight, rectangular grid pattern of streets, broken only by a few large interior open spaces and a parade ground.[26]

The influence of the Randall Plan on the Manhattan waterfront has not been adequately recognized. The plan created a rigid pattern for further east-west expansion of the island. To make rectangular plots that would be easy to develop and could earn the greatest profit, paper streets straightened every jagged piece of land along the water, even if, due to the vagaries of topography, it could not be reached. Once again, by drawing a line between points (this time along the four-hundred-foot boundary) West Street was extended to Canal Street, Twelfth Avenue created north of West 34th Street, and Avenue A drawn to connect East 55th Street to East 17th Street.[27]

No parks were provided along the rivers; instead, the tight development of narrow finger piers at each street-end south of West 59th Street and Corlears Hook was encouraged. In all the map set 155 east-west streets two hundred feet apart, providing direct approaches to piers that could be built at the ends of these roadways.[28] Ideally the maximum amount of waterfront could be developed for commerce, and access to piers projecting on all rivers, from the Battery to Marble

Hill, would be easy. However, as traffic increased the thoroughfares became congested, and as ships became longer and wider, the narrow piers became obsolete.

Port of Commerce and Entry, 1807–1850. During the first half of the nineteenth century, Manhattan again burst beyond its formal limits. The phenomenal prosperity brought on by an explosion of New York's maritime economy caused intense pressure. Between 1820 and 1830, its export and import values doubled, and the city surpassed its rival ports, Philadelphia and Boston.[29]

Room was needed to build new carriers, so land was pushed out into the East River above Corlears Hook. In 1824 Henry Hall established the country's first dry dock at East 10th Street; nearby were the bustling East River shipyards. Soon, such ancillary activities as machine works, sail makers, ships' carpenter tool manufacturers, iron and brass foundries, and mahogany yards lined the waterfront above Corlears Hook from Grand to East 12th Street.

The production of larger vessels by local shipbuilders, the introduction of the steamboat in 1807, and the opening of the Erie Canal in 1825 created demands for longer piers in deeper water and for all sorts of new landings. Down the launchway came the 319-ton, ninety-eight-foot-long *Savannah,* soon superseded by the *Reglus*—twice as heavy and half again as long.[30] Seeking berths with these ships were the traditional barks, brigs, schooners, and sloops, and newer ocean-going sailing packets, coastal and river paddlewheelers, sail and steam-driven ferries, canal barges, sail-powered lighters, and towboats. By 1839 the shores of lower Manhattan had acquired their sawtooth appearance. Finger piers extended along West Street from every street-end between Vesey and King streets and along South Street from the Battery to Catherine Street.[31]

A population surge also caused changes in the city's physical profile. The Irish, English, and French swelled the census in the 1830s from four thousand to thirty-two thousand. This figure doubled in the next decade.[32] Because land on the small island was finite, the city was forced to move northward and outward to accommodate the new New Yorkers. The uptown city limits moved above Duane Street in 1807 and to Washington Square thirty years later.

Encouraged by the New York State Legislature's 1807 extension of the four-hundred-foot-line northward, from Corlears Hook to East 40th Street and from Spring Street to West 70th Street, river edges ballooned. In the East River shipyard district above Corlears Hook, Tompkins Street replaced East Street, and rickety housing for immigrants was erected on the intervening filled-in blocks. Here lived carpenters, calkers, joiners, sawyers, and seamers.

On the Hudson in the 1830s water lots were made into land to the west of Tenth Avenue, some as far north as West 16th Street. Fur trader-turned-realtor John Jacob Astor contributed to this expansion, filling in his water lots between West 14th and 16th streets well beyond Tenth Avenue. Shorefront bluffs were leveled to make streets and land for factories, warehouses, and row houses. The debris was simply dumped into the river toward newly mapped Thirteenth Avenue.[33]

Urban Refuse, 1800–1862. As New York grew larger, material for landfill was plentiful. On the route of the city's northward movement, with the help of new blasting materials and steam-powered equipment, more hills were leveled, streets and foundations were excavated, and old buildings demolished. All of this activity created large amounts of wood, sand, rock, and topsoil that had to be removed. It was estimated in one report that over one thousand loads of dirt could emerge from digging a single cellar.[34]

When fill was plentiful the city made a modest profit, charging cartmen from three to five cents for the privilege of dumping each load. To keep carting costs down, debris was usually deposited at the nearest low ground. Thus, as in Boston and San Francisco, the sand from inclines at Burnt Mill Point found its way to the East River swamps below 14th Street, from Avenue B to the water, and the dirt from three hills that were lowered in grading West Side avenues helped to fill in a cove between West 21st and 24th streets. Work relief projects to combat periodic depressions often provided labor.[35]

Disaster was also a source of landfill. In 1835, after fire devastated much of lower Manhattan, the prominent landowners to the west of Bowling Green finally petitioned for the extension of West Street southward from Cedar Street to the Battery. Their rationale for this improvement was that "the great accumulation of earth and refuse material growing out of the late fire renders the present a propitious moment to effect this desirable object."[36]

Clean landfill unearthed in building activity was most sought after, as it caused fewer health and settlement problems. It was also often scarce. In 1823, when the construction industry had siphoned off materials such as clay, sand, gravel, and rock to use in mortar and building foundations, the remains of inclines below East 14th Street were barely enough to fill in new bulkheads and outer streets on the Lower East Side. Additional stone was poured into the neighborhood from as far away as Vermont. As an incentive, the city was then forced to pay the cartmen anywhere from twelve to twenty cents.[37]

Despite the scarcity of clean fill, as people crowded into Manhattan and real estate prospered, expansion of the edge continued and the Street Superintendent orchestrated this outward growth. (Map 4) Beginning in 1802 with the formation of the Department for Cleaning Streets, he took charge of trash disposal. His job was the removal not only of household refuse but cinders, street sweepings, and animal dung. He also designated inland city dumps where the refuse would rest as fill or be sold as fertilizer. More importantly, he selected (often with the approval of aldermanic committees and members) the wharves and piers where excess material would be loaded into scows for shipment out of the city or to sea. And here is where the city once again burst beyond its legal limits.[38]

The berths for garbage and manure scows that toted refuse out of the city were called dumping boards. They were rude platforms at the river's edge, above the level of the street. (Figure 1) Here the contents of carts hired by the city or independent collectors were emptied and eventually pushed overboard into the troughs of waiting scows. In the early nineteenth century the dumping boards were located on rapidly expanding edges of Manhattan, at Whitehall Slip next to the Battery, at James Slip where the construction of South Street was under way, and at Park Place on the Hudson, two blocks above Washington Market.[39]

With more people and better collection methods, the city's garbage supply grew plentiful. Manhattan's population once again doubled between 1840 and 1860. In 1856 over 107,000 loads of garbage and refuse were removed from its streets. By 1860 this number had risen sevenfold to 798,000.

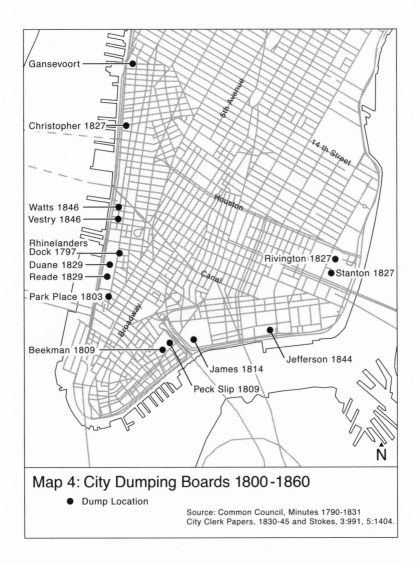

Map 4: City Dumping Boards 1800-1860

● Dump Location

Source: Common Council, Minutes 1790-1831
City Clerk Papers, 1830-45 and Stokes, 3:991, 5:1404.

The problem was where to put the refuse once it was collected. A portion of the city's waste was used to fill the few remaining swampy areas, and a considerable amount was taken by scow (at 5¼ cents per cubic yard) to Blackwells Island, the city's first major, organized landfill project. Some

Figure 1. Dumping Board, West 30th Street, 1910. (Courtesy New York City Municipal Archives).

materials were shipped to sea, but this was an expensive method of disposal costing as much as fourteen cents per cubic yard.[40]

The most profitable disposal sites were the dumping boards. On the West Side, between Laight and Hubert streets, and at Watts, Christopher, and Gansevoort streets, ash, offal, and garbage were allowed to collect and spill into the slips. It was cheap fill—in fact, it was free—and there were few complaining neighbors to cause its removal. Soon water was turned into land for markets and peculiarly shaped structures for shipping. By 1862, the equivalent of fifty-six twenty-five-by-one-hundred-foot lots had been created near the

46

dumping boards. All of them were beyond the city's four-hundred-foot limit, the state bulkhead lines, and the outer streets of the Randall Commissioners' Map.[41]

Washington Market, 1812–1862. The potato crop failure in Ireland, the shift from sail to steam in transatlantic passenger shipping, the discovery of gold in California, and construction of the world's longest and fastest sailing clippers saw a burst of maritime activity never before experienced in New York. The shape and use of Manhattan's West Side was particularly affected. Longer ships, finding little room to maneuver in the East River, moved to already crowded piers on the Hudson. Shipbuilding, repairs, and other services requiring river frontage followed their source of business, settling in undeveloped sections in Brooklyn and above West 23rd Street.

In 1855 a New York Harbor Commission was appointed by the state to investigate the condition of New York's waterways and to prevent "extensive and alarming" encroachments into the harbor.[42] Among the properties affected by the deliberations of this commission were the waterfront dumping sites and Washington Market. The landfill related to these activities had created valuable parcels of real estate, the ownership of which was suddenly in question.

The event that precipitated the commission's appointment was expansion of one of Manhattan's West Side public markets. The parcels, which the city had kept in reserve during the years that it sold most of its shore properties, or changing its policy, had repurchased, had been developed to provide income for the municipality. Rental, wharfage, and other fees from public markets, piers, and dumps helped to defray the costs of their operation. An even more lucrative source of

47

revenue was the lease and sale of lots. This provided an additional incentive to move these municipal facilities further into the rivers, and the city, acting like a private entrepreneur, did just that.

The two blocks between Dey, Fulton, and Vesey streets, Washington to West Street (today the World Trade Center), were part of a sixteen-parcel eighteenth century public landfill. This property again pushed westward shortly after the beginning of the War of 1812. With the growth of Manhattan to the north, Greenwich Street had become an "important" thoroughfare and plots were in demand, particularly for elegant dwelling houses. Having sold several lots to pay back municipal debts incurred during the Revolution, the city now owned seven lots fronting on this popular street, all occupied by an aging, overcrowded "Hudson" Market.[43]

To provide a better place for people to buy food, and to capitalize on the real estate potential of the market lots, the city created new land. Between 1813 and 1816 the Greenwich Street plots were sold for approximately $5,600 each. The proceeds were used to build a modernized and renamed "Washington" Market on fill deposited in the slip between Fulton and Vesey streets, to the west of Washington Street.[44] (Map 3)

Coincidentally, the block to the south—from Dey to Fulton streets, Washington to West Street—known as Varick Basin, was acquired by the city from its owner, former Mayor Richard Varick. This too was filled in and upland lots were sold at prices that did not quite offset the filling and purchase costs.[45] The waterfront strip along West Street was retained to expand an existing public dock. With city ownership of the edge, the two blocks from Dey to Vesey Street were ripe for expansion westward.

The market's success between 1844 and 1847 caused more fill to be deposited. Land was made for new sheds, called

48

"West Washington" Market, at a cost of four thousand dollars. These structures were immediately rented at a one thousand dollar profit. Then dozens of wooden shanties, selling country produce and each paying from three to four dollars a day, were moved out even farther onto fill.[46]

With the promise of ever greater returns, it was suggested that the thirty-year-old market on Washington Street be reconstructed. What better site than the two-hundred-foot bulge of unfinished land that had formed beyond the often-straightened outer limits of the city. Again, the proceeds from the sale of the more valuable city land between Washington and West streets would defray the cost.

In the efforts to implement this development, the very laws that had been created to expand the ground now became constraints. First to be questioned was the Outer Streets and Wharves Act, which in 1837 had established West Street to remove the natural and artificial indentations along the lower, western edge of the city. In 1849 Mayor William F. Havemeyer used the act to veto the proposal to build a new market on the fill beyond West Street, arguing that it would destroy the integrity of the now-straightened building line. If the city were to expand beyond West Street, warned Havemeyer, it must not do so in a piecemeal manner.

He called for adoption of a "proper and general plan . . . of improvement along the Hudson," which would replace any lost piers. If a new exterior street were needed, Havemeyer wanted the "best arrangement for commercial [meaning maritime] accommodations."[47] This "plan" was typical of appeals for improving the waterfront that would recur into the twentieth century, in which the port was of paramount concern. There was no mention in Havemeyer's statement of the municipal services that might be displaced by private development, nor of creating an orderly blueprint for landfill.

As discussion on the expansion of Washington Market proceeded it became clear that the real concern was neither the
city service nor the jagged edge. Rather, the issue was the
adequacy of the city's piers. While Manhattan was rapidly
expanding into the Hudson River to provide revenues to build
new market space, there were complaints that reduced shipping facilities threatened the very heart of the city's economy.
A report to the State Senate in 1854 described the sad plight
of one vessel, the *Great Republic:*

The harbor of New York, at the close of last year, was graced with
the presence of one of the noblest ships that ever floated upon the
sea . . . designed, if successful, as the pioneer of a line of vessels to
trade between the new world and the old. The supposed capacity
of the Great Republic was equal to a freight of forty thousand barrels
of flour, or more; but owing to the difficulties of egress from the
port, the necessary care of the ship in the harbor, the narrowness
of the river . . . the owners of the Republic . . . deemed it unsafe to
load this noble vessel with more than three-fourths of a cargo. The
Republic . . . lies a wreck at one of the docks of New-York.[48]

In short, as land pushed out and new piers were built further
into the crowded harbor to accommodate two to five thou-
sand-ton steamers, the waterways were becoming narrower
and shallower and more dangerous.[49]

New York State still held title to the lands beyond the
four-hundred-foot mark, and, since completion of the Randall Commission Map, had exclusive power to change exterior street lines. With this authority, and through The New
York Harbor Commission, it placed its imprint on the shape
of Manhattan's waterfront. Charged by the state with rec-
tifying the problems hampering shipping, the commission
recommended permanent pier and bulkhead lines in 1856.[50]

Although the concept of restraints was new to a city that

had heretofore enjoyed unbridled expansion, the new borders merely circumscribed what existed. They more or less co-incided either with the now filled-in four-hundred-foot lines that had been prescribed in the Dongen and Montgomerie Charters, or with the piers and bulkheads as found in 1855. The boundaries for the market parcels on the Hudson from Dey to Vesey Street were more vigorously drawn. Here, the Harbor Commissioners claimed that landfill had "deprived the port of dock accommodations in the center of business" for six to ten steamers.[51] Consequently, they defined a permanent bulkhead line that made Washington Market an illegal encroachment in the river.

At issue too was money. The new bulkhead limits reinforced a state claim to ownership of the new-made land beyond Greenwich Street for which the city had never purchased water rights. There were protracted legal wranglings, and in 1861, four years after publication of the Harbor Commissioners' *Report,* the city bought from the state the eighty-three lots that it had illegally created between Dey and Vesey streets. At $150,000 this was a Pyrrhic victory for the state, for it was one-third of the cost of the Varick property purchased by the city a half-century earlier.

In the final analysis not legal restrictions, but delay of fourteen years in establishing a new market, effectively put a halt to landfill expansion on the Lower West Side. During this time other interests became paramount. Desirous of keeping the waterways clear, the shipping industry had helped in the establishment of the State Pier and Bulkhead Lines, beyond which no further construction could take place. In addition, the Hudson River Railroad Company, incorporated in 1846, now operated an important freight and passenger line along Tenth, Eleventh, and Twelfth avenues. When the Spuyten Duyvil railroad bridge opened in 1853 trains were able to run from Chambers Street all the way to Albany, and this became the only main line to serve the West Side docks that did not require water transfer.[52]

With all of this activity, by 1860 residents of the city were moving north of 34th Street, leaving the downtown waterfront to be the center of the country's prime port. Ironically, the market activities would eventually be pushed back onto Washington Street, and the landfill removed in order once again to elongate the piers. The nineteenth-century land explosion at the Lower West Side had ceased.

NOTES

1. Writers' Program of the Work Projects Administration for the City of New York, *A Maritime History of New York* (Garden City: Doubleday, Doran, 1941), 5–6. (Hereafter abbreviated WPA, *Maritime History*.)
2. Richard Baiter, Office of Lower Manhattan Development, *Lower Manhattan Waterfront*, June 1975, 5.
3. Robert Greenhalgh Albion, *The Rise of New York Port, 1815–1860* (New York: Scribners, 1939), 20. For maps showing the original high-water line and subsequent landfill see: New York City Office of Lower Manhattan Development, Department of City Planning, *Lower Manhattan Waterfront* (New York, June 1975), 4; City of New York, City Planning Commission, *The Waterfront: Supplement to Plan for New York City* (New York, January 1971), 106; J. W. Gerard, Jr., *A Treatise on the Title of the Corporation and Others to the Streets, Wharves, Piers, Parks, Ferries and Other Lands and Franchises in the City of New York* (New York: Baker, Voorhis, 1873), endplate; I. N. Phelps Stokes, *The Iconography of Manhattan Island 1498–1909*, 6 vols., 1915. Reprint (New York: Arno Press, 1967), 3:Plates 174–80; "Map of the City of New York showing the original high water line," in D. T. Valentine, ed., *Manual of the Corporation for 1852* (New York: McSpedon and Baker, 1852), 462; "Plan of the City of New York showing the made and swamp land," in *Valentine's Manual for 1856* (New York: McSpedon and Baker, 1856), 202; Egbert L. Viele, "Sanitary and Topographical Map of the City and Island of New York," in Citizens Association of

New York, *Report Upon the Sanitary Condition of the City* (New York:
D. Appleton, 1866), frontispiece; and Bailey Willis and R. E. Dodge,
"Physiographic Features of the District," *New York Geological Survey,*
1901, Map Room, File "N.Y.C. Folio 1902, No. 83," New York
Public Library.

4. The Hudson Ship Canal separated this peninsula from the Bronx by
cutting through Manhattan at 220th Street and enlarging the creek at
the southern end of Marble Hill. Now there were two islands at Manhattan's northern tip. In the twentieth century the Harlem River was
filled in around the northern edge of Marble Hill, attaching this island
to the Bronx. The improvement of 215th Street made the Spuyten
Duyvil peninsula part of Manhattan. M. Dripps, "New York City,
County and Vicinity," in *Valentine's Manual for 1866* (New York:
Edmund Jones, 1866), frontispiece; Haagstrom, *Atlas: City of New
York,* 1976 (18th ed.), 3–4; "Map of the Kings Bridge Section copied
from the Preliminary Map of the Commissioner of Washington
Heights," November 1860 (attributed to Frederick Law Olmsted),
Map Room, File "Kings Bridge," New York Public Library.

5. Henri and Barbara van der Zee, *A Sweet and Alien Land: The Story of
Dutch New York* (New York: Viking, 1978), 6; and Stokes, *Iconography,*
6:Plates 84 B–g, and 3:Plates 174–80.

6. WPA, *Maritime History,* 32–33.

7. Dongen Charter, 1686, Sec. 3, 14. Exception: Battery, Governor's and
Queen's Garden—Crown property. See Landfill maps in Gerard, *Treatise on the Title,* and Stokes *Iconography.*

8. WPA, *Maritime History,* 61; and Claudia Lorber, "Digging Up Our
Urban Past," *New York Times Magazine,* 12 April 1981, 54–56, 68,
87–88, 120–22. See also the following articles in *Seaport,* Fall 1980:
Susan Kardas and Edward Larabee, "Landmaking in Lower Manhattan—When New Yorkers want more land they just make some," 15–
19; Lorber, "We've Been Digging the Shores of Dutch New York—
Pre-skyscraper digging in the 1643 Stadt Huys site," 8–14.

9. "New Yorke, 1695," in Thomas A. Janvier, *In Old New York* (New
York: Harper and Brothers, 1894), 24. Janvier has an excellent set of
maps showing the growing waterfront. One must exercise caution in
their use, however, as they often show anticipated or legislated, rather
than de facto change.

10. WPA, *Maritime History,* 61.

11. Citizens Association, *Report,* 14.

12. Burgis View of the "South Prospect of the flourishing City," 1717,
in Stokes, 1:239–51, Plate 25.

13. *Minutes of the Common Council of the City of New York, 1760–90,* 8 vols.

(New York: Dodd, Mead, 1905), 2:93, 127, 134, 138; Stokes, 4:429. Backgrounds on individuals were compiled from William Thompson Bonner, *New York: The World's Metropolis 1623-4—1923-4* (New York: R. L. Polk, 1924).

14. Stokes, *Iconography*, 4:453; "Plane of New York in 1729," Surveyed by James Lynne, in Janvier, *In Old New York*, 36, 37.

15. Janvier, *In Old New York*, 36-37.

16. Charles Lockwood, *Manhattan Moves Uptown* (Boston: Houghton Mifflin, 1976), 33.

17. Title Guarantee and Trust Company, Title Searches, Blocks 17-19; Rufus Rockwell Wilson, *New York: Old and New. Its Story, Streets and Landmarks* (Philadelphia: J. B. Lippincott, 1903), II:11-30; 55-79. The Delancey land was confiscated and sold to Nicholas Roosevelt. After the Montgomerie Charter granted the four-hundred-foot extension from Corlears Hook and from Charlton Street south to the Battery, the city once again sold or granted perpetual leases to the new underwater land.

18. Murray Hoffman, *Treatise Upon the Estate and Rights of the Corporation of the City of New York as Proprietors* (New York: McSpedon and Baker, 1853), 188.

19. WPA, *Maritime History*, 42, 60; Albion, *Rise of N.Y. Port*, 305; WPA *Boston Looks Seaward*, 43; and "Plan of the City of New York 1767" Surveyed by Bernard Ratzen, in Janvier, *In Old New York*, 48.

20. WPA, *Maritime History*, 65-76.

21. Ibid., 90.

22. George Ashton Black, "The History of Municipal Ownership of Land on Manhattan Island," *Columbia University Studies in History, Economics and Public Law* 1 (1897):36.

23. Graphic evidence of this may be seen in the "Plan of the City of New York" from an actual survey by F. Maerschalk, in Janvier, *In Old New York*, 39.

24. Gerard, *Treatise on the Title*, 39, 35; and Hoffman, *Treatise Upon the Estate*, 33.

25. Hendrik Hartog, *Public Property and Private Power: The Corporation of the City of New York in American Law, 1730-1870* (Chapel Hill: University of North Carolina Press, 1983). The city, posits Hartog, in placing restrictive covenants on private individuals who purchased water lots, engaged in formal planning for its commercial expansion. When increasingly the covenants failed to create an ordered commercial edge, the city sought new powers of implementation from the state.

26. William Bridges, *Map of the City and Island of Manhattan with Explan-*

atory Remarks and References (New York: T. Swords, 1811). The area unaffected by the map was south of Gansevoort Street on the west and south of Houston Street on the east.

27. Ibid., 24.

28. Streets above 155th Street were laid out by the Parks Commissioners after 1851.

29. Albion, *Rise of N.Y. Port,* 389, 392–93.

30. John H. Morrison, *History of New York Shipyards* (New York: William F. Sametz, 1909), 23–51, 150; A. A. Rikeman, *The Evolution of Stuyvesant Village* (Mamaroneck: Curtis G. Peck, 1899), 74; and Title Guarantee and Trust Co., Title Searches, Block 367.

31. "Plan of the City of New York Showing Fire Districts, etc.," Narine and Co., 1839, Map Room, File "1830," The New-York Historical Society; Albion, *Rise of N.Y. Port,* 303–16; and Morrison, *N.Y. Shipyards,* 54.

32. Albion, *Rise of N.Y. Port,* 337–38.

33. City Clerk Approved Papers, File "Streets, 1830–7," New York City Municipal Archives, (hereafter abbreviated City Clerk); "Map of 1836," George W. Smith Surveyor, Astor Papers, File "Maps 1836," Manuscript Collection, The New-York Historical Society; Frank B. Kelly, *Historical Guide to the City of New York* (New York: Stokes, 1913), 115; and Stokes, *Iconography,* 5:1938.

34. Karadas and Larrabee, *Seaport,* Fall 1980: 17; and New York City Board of Assistant Alderman, Doc. 9, 9 October 1848, 120.

35. Citizens Association, "Reports of the Sanitary Inspectors," *Report,* 1–348; *Minutes of the Common Council of the City of New York, 1789–1831,* 19 vols., New York, 1917–18, costs for landfill are listed under expenditures at the end of each year 5–19; and Stokes, *Iconography,* 5:1463–1507.

36. Petition of Van Cortlandt, White, Richard and Robert Morris, and Edgar, 9 January 1836, City Clerk, File "Streets, 1836."

37. Citizens Association, *Report,* 97–98, 121–22; Rush C. Hawkins, *Corlears Hook in 1820* (New York: J. W. Bouton, 1904); City Clerk, File "Streets, St-W, 1823–46" and "Tompkins, 1826"; Assistant Aldermen, Doc. 9, 9 October 1848, 120; *Annual Report of the Department of Docks* for the years 1870–1890; and City of New York, Department of Docks, Letters, New York City Municipal Archives. (The Department of Docks is hereafter abbreviated DD.) Costs for carting were found scattered among these letters. In addition, garbage, and refuse tonnage and carting costs are found in *Annual Report of the City Inspector of the City of New York,* 1857–73.

38. Eugene Moehring, *Public Works and the Patterns of Urban Real Estate*

Growth in Manhattan, 1835–1894 (New York: Arno Press, 1981), 53; and *Ordinances of the Mayor and Aldermen of the Communality of the City of New York*, ch. 11, 14 March 1839. Street cleaning tasks were assigned to the following agencies in the nineteenth century:

The Department for Cleaning Streets, 1802–65;

The Board of Health, 1865–73;

The Police Department, 1873–81;

The Street Cleaning Department, 1881–98.

39. George E. Waring, Jr., *Street Cleaning and the Disposal of the City's Waste: Methods and Results and the Effect Upon Public Health, Public Morals and Municipal Prosperity* (New York: Doubleday and McClure, 1898), 208; and City Clerk, File "Streets, 1835." The location of nineteenth-century dumping boards was reconstructed from: Common Council, *Minutes, 1790–1831;* City Clerk, File "Wharves, Piers and Slips, 1830–45"; and Stokes, *Iconography,* 3:991, 5:1404.

40. *The Annual Report of the City Inspector,* Board of Aldermen, Doc. 19, 21 September 1857; *City Inspector,* for year ending 31 December 1860; *City Inspector,* Doc. 5, 17 January 1861; and Waring, *Street Cleaning,* 210.

41. *Report of the Commissioners of the Land Office Relative to New York Harbor Encroachments,* Sen. Doc. 10, 19 January 1862, 74–75.

42. New York State, *Report of the New York Harbor Commission of 1856 and 1857* (New York: C. S. Westecott, 1864), preface.

43. Lockwood, *Manhattan Moves Uptown,* 42; and Thomas F. DeVoe, *The Market Book Containing a Historical Account of the Public Markets in the City of New York* (New York: printed for the author, 1862), 324.

44. Black, "History of Municipal Ownership," 36. Washington Market was also known as "Oswego," "Bear," and "Hudson's" Market.

45. The city's costs according to Black, "History of Municipal Ownership," 45, 50, were as follows:

$ 65,000 Underwater land Dey St.

27,000 Bonds

47,000 Fill and bulkhead

$139,000 Total

The lots were sold for $127,000 or at an 8 percent loss.

46. DeVoe, *Market Book,* 441. How much of this accrued to the city is unknown. Validated leases for stands in the market shed were two to ten dollars per week.

47. Ibid., 446.

48. State of New York, *Report No. 49,* 10 February 1854, 13.

49. N.Y.S., *N.Y. Harbor Commission,* 17–18.

50. Ibid., 1.
51. Ibid., 129.
52. Carl W. Condit, *The Port of New York: A History of the Rail and Terminal System from the Beginnings to Pennsylvania Station* (Chicago: University of Chicago Press, 1980), 32–35.

CHAPTER 3

The Master Plan

As we have seen, between 1647 and 1861 the waterfront of Manhattan developed in a fragmentary way, primarily according to individual needs and resources. By the last quarter of the nineteenth century, however, the very success of the port caused problems that threatened not only local owners and users but the economy of the city and the state.

By 1870 the port of New York had achieved international preeminence. Over ten thousand vessels berthed here (thirty times more than a century earlier), carrying a total tonnage in foreign, coastal, and domestic trade that was twice that of London. In fact, New York's maritime activity surpassed the combined movements of that European capital and its rival, Liverpool. According to the *Real Estate Record and Builders' Guide,* New York was "destined . . . beyond doubt at no very distant day to become the center of the trade and commerce of the world."[1]

This rosy future, however, was threatened by the "dis-

graceful system of wharves and piers which characterize[d]"
the city at that time.[2] While the East and Hudson rivers were
solidly lined with docks from Whitehall to East 23rd Street
and Morris to Morton Street, a distance of over 5½ miles,
berths were scarce. Canal boats, limited to five piers on the
East River between Moore Street and Cuyllers Alley, were
begging for space occupied by coastal steamers. The largest
ships in the port, steamers traveling to Europe and the Orient,
vied with each other for slips on a meager four blocks of the
Hudson between Canal and King streets. Among those also
bidding for landings were local river and Sound steamers,
pleasure craft, produce boats in the West Indian trade, ice,
coal, stone and lumber barges, and garbage scows.[3]

New industries were also entering the fray, using piers
formerly occupied by shipping and establishing footholds on
the undeveloped blocks to the north. After 1866, in the absence of bridges or tunnels, railroad campanies added to their
fleets floats capable of carrying entire cars packed with goods
across the river. Lighters and car floats lined up at old piers
between Morris and Rector streets and at new ones constructed between West 30th and 33rd streets on the Hudson.
Also, gas companies introducing electricity to the city were
building landings for coal barges and utility floats from East
11th to 16th streets, West 16th to 20th streets, and West 44th
to 46th streets.

Not only were there insufficient landings, but those that
existed were in deplorable condition. Of the approximately
126 piers in the East and Hudson rivers (down by a third
since 1864), only one was classified by the city as "very
good." One-sixth of the structures were "good," the pre-
ponderance of which were newer piers on the Hudson north
of Canal Street. The remaining one hundred or so structures

suffered a myriad of problems: they were too short, under-pinnings were gone, decking was missing, and whole portions of docks and bulkheads had rotted away to nothing.[4]

Shallow water was a pervasive problem, for the river edges had become permanent sumps for New York's sewage. As the population grew and running water was introduced into homes, ducts were constructed under east-west streets. These pipes carried household and gutter debris to outlets in the banks of the rivers.[5] The finger piers, which projected at right angles from the land into the water (either over or adjacent to these outlets), prevented the free flow of the tides from carrying the sewer dirt into the open sea. Instead, the piers formed solid barriers that encouraged an annual pile-up of as much as eighteen inches of muck.[6]

In the past, it had been more profitable to ignore the refuse than to remove it. For example, the early nineteenth-century Washington Market expansion was helped when the water between Vesey and Dey streets was allowed to fill with silt and garbage because dredging costs were as high as the selling price of a single lot.[7] After the establishment of the State Harbor Lines in 1856, the slips that remained in maritime use continued to silt up and wharf owners were forced to pay up to 30 percent of the annual income of a pier for dredging.[8] By 1870 the land-end portion of many piers on the East and Hudson rivers was accompanied by a scant zero to seven feet of water. Considering that the bottom of Robert Fulton's *Clermont* had sat seven feet into the water when empty, this left no room for docking larger contemporary craft.

Problems were also legion above ground. Loading and unloading, performed by hand (despite the availability of steam-powered hoisting cranes) was painfully slow. Because the majority of piers and bulkheads were open and unprotected, merchandise, waiting days or months to be moved,

Figure 2. Piers 46–49, West 6th to West 9th Street, Hudson River, 1903. (Courtesy New York City Municipal Archives).

was stolen and damaged. (Figure 2) Delivery and distribution of goods were blocked by horse-drawn railroad cars and by carts backed up on narrow, dirty streets. (Figure 3) Support services were dispersed, with warehouses, insurance brokers, and repair yards often located miles from those landing facilities that were in decent order.

All of this resulted in costs to those using the port as well as to the state, the city, and to public and private dock owners. The former had expenses in waiting, extra haulage, and property loss that were estimated at five dollars per ton. The latter lost revenue in forfeited wharfage fees and more importantly, from the shipping that moved to more compatible docks in New Jersey.[9]

Figure 3. Congestion on West Street, c. 1905. (Courtesy New York City Municipal Archives).

Reform, 1860–1870. Efforts to fix New York's ailing port facilities were mired in a labyrinth of agencies. Repairs and building of new piers were under the jurisdiction of both the Bureau of Wharves, Piers and Slips of the City Street Cleaning Department and the State Board of Commissioners of Pilots. The latter could build and repair if the city failed to do so, while the former depended on periodic state bond issues (often vetoed by the Governor) and a three-quarters vote of the city's Common Council for any significant funds.

Rents collected from vessels while berthed also paid for pier improvements, but here too the lines were tangled. The state set the wharfage rates; the city and individual dock owners collected the fees. Owners complained that the state-set rates had little relation to their maintenance and improvement expenses. Meanwhile, the city receipts were not allowed to be used to add new planking to a broken-down deck, but were contributed to interest payments on the city debt.

Even the administration of the port was bifurcated and confused. City Inspectors of Incumbrances on the Wharves, and a Common Council Committee on Wharves, Piers, and Slips performed many of the same tasks as did the state appointed Captain of the Port, with his eleven Harbor Masters. The latter assigned berths to vessels needing space, collected fines from boats which failed to debark on time, and issued summonses when goods lingered on the piers and bulkheads.[10]

Efforts to escape from this morass resulted in memorials, reports, and surveys, all of which recommended the creation of an independent body of waterfront overseers. State Commissions and the sale of every pier in the city to the highest private bidder were suggested. In 1865, the New York Chamber of Commerce recommended the incorporation of a quasi-private group, the New York Pier and Warehouse Company, to oversee the installation of an elaborate complex of piers and warehouses that would extend along West Street on the North River from the Battery to Central Park.[11]

Several years later business sentiment, although still ambivalent, leaned toward city control. The Citizens Association, concerned with municipal corruption and declining services, devised a plan to change the way the city was managed, and docks were a subject of their reform. In 1869 the Association, comprised of large property owners in the city and backed by those with interests in the port (including the Chamber of Commerce, the Shipowners' Association, the Produce Exchange, and banking houses and manufacturing firms), recommended creation of a thirteen-member City Dock Board.[12]

The City Charter of 1870 finally mandated a formal mechanism, totally within the city government, to address the physical problems of the waterfront. Included were a permanent body of overseers, legal recourses, and money. A municipal Dock Department was created. It was composed of administrative staff and a Dock Board with five commissioners; it had "exclusive charge and control" of all waterfront property—slips, piers, bulkheads, and structures thereon—belonging to the city then, or to be acquired in the future. To solidify this absolute dominion, the state deeded to the city all remaining ungranted lands under water encircling Manhattan. In addition, the board was given the authority either to buy or to initiate proceedings to appropriate any wharf property not publicly owned. The board's broad powers were restrained, however, by the financing section of the charter. This required the board to apply to the City Comptroller to issue dock bonds, the sale of which would pay for all expenses. The annual limit on the issuance of these bonds was three million dollars.[13]

The initial Dock Commissioners appointed in 1870 were prominent business and professional men. Their backgrounds were reflective of the problems that existed and presaged the type of development that would ensue. Board President John T. Agnew was a tobacco exporter, while Wilson G. Hunt, a dry goods merchant, was financier of the first Atlantic cable. This invention made obsolete a precarious network of pick-up boats and carrier pigeons that in the early nineteenth century rushed news from incoming steamers to the local press.[14] Two members of the Dock Board were experts in finance and law: William Wood, a banker who also served on the Board of Education, and Richard Henry, an attorney. The fifth commissioner, Hugh Smith, represented ground transportation, a major competitor for waterfront space. His Madison Avenue Stage Company was one of several horse car concerns that provided the only form of public transit along the city's increasingly congested streets.

Although the Dock Board was composed of laymen, a professional staff was in charge of the physical improvements. As Engineer-in-Chief of the Department of Docks the Board hired George B. McClellan, whose prior career was evidence of the importance of the job. "Little Mack" had held or sought the most prestigious positions in the United States; he was a Civil War general and a candidate for President. A graduate of West point, he was trained in nineteenth century warfare, which included constructing roads and bridges in order to move troops to remote battlefields. McClellan perfected his engineering skills in surveying the route for the Northern Pacific Railroad across the Cascade Mountains.

Called to active duty in the Union Army during the Civil War, he ejected the Confederates from the western part of Virginia and quickly rose from Major General to General-in-Chief in command of the Union troops around Washington. Despite controversy over his leadership, McClellan was victorious at Antietam, checking Robert E. Lee's first invasion of the North.

When removed from his army command at the age of thirty-six, McClellan tested his potential for civilian leadership. In 1864, as the youngest candidate to date, he ran for the presidency and lost to Abraham Lincoln. After a period of semi-retirement, McClellan was hired to perform engineering duties for which he had been trained—directing the New York City Dock Department's program of improvements.[15]

As a person McClellan brought a renown to the department that would be useful in giving credence to proposed changes;

however, his profession was common to waterfront development. For over a century, engineers had held sway over the growth of important port cities in Europe because they had the training to cope with problems that were particular to shorefront expansion. They could test materials scientifically (a new bulkhead in New York required durable cement), estimate construction costs accurately (a chronic problem for the city), and coordinate several modes of transportation.[16]

Outline and Sketches, 1870. Included in the chapter of the 1870 City Charter that created the Dock Department was a section on waterfront planning. For the first time in the city's history, information was to be gathered, ideas proposed, and a final, long-range plan delineated. This document was intended to be the official guide for any future modification of the water's edge. Under the prescribed planning process the Dock Board was instructed to hold public hearings and citizens were invited to submit waterfront improvement plans. Unlike the competition for the design of Central Park, in which a winning architect was selected, the Dock Commissioners were given the latitude to choose features from select proposals or to create an entirely new scheme of their own.

There was no lack of information to choose from. Between 23 June and 31 December 1870, seventy maps, drawings, and remedies were presented to the Dock Board orally and in writing. The suggestions ranged from methods of protecting wooden piles from toredos—destructive underwater creatures, which thrived in the less polluted waters of the time—to megastructures answering urban needs for improved transportation, storage facilities, markets, sewer and power lines, and parks. Ostensibly, the proposals addressed problems associated with New York's waterfront commerce. They were also a microcosm of major late-nineteenth-century physical problems that were inexorably tied to life in the interior of the city.[17]

Because a good pier system was important to the prosperity of New York, the majority of the plans presented to the Dock Department offered designs to replace the tumbledown fringe. The simplest showed rows of attractive new finger piers of undesignated materials. Others anxious to replicate the magnificent marine works of Europe, some of which dated back to ancient times, as well as those newer installations closer to home in Boston, Buffalo, and Albany, pressed for new materials and elaborate storage spaces. Captain James A. Nichols, a shipmaster who had traveled extensively, believed the city should abandon wooden piers in favor of thirty new, longer-lasting iron structures. W. H. Emmons and F. Kissam proposed an array of hollow stone piers encircling Manhattan, on top of which would sit a continuous, one-story warehouse with railroad tracks on its roof; underneath would lie additional storage space. The warehouses envisioned in the wet basin plan of William C. H. Waddell were multistoried, incorporating spaces for marine services on the first level with passenger and freight depots above.[18] (Figures 4, 5)

Sanitation was another prevalent theme. Outbreaks of yellow fever, cholera, typhus, and smallpox had periodically decimated the city's population since 1791. In 1857 twenty-seven out of every one thousand New Yorkers died of disease. This compared unfavorably with a nationwide ratio of fifteen-to-one-thousand. While a smallpox vaccine was available, the other illnesses were yet to be controlled. Theories of what caused these deadly fevers and intestinal infections were plentiful. For some who found nasty outbreaks in the slums, the cause was moral turpitude. For others, a vague,

Figure 4. Elevation of Docks and Warehouses as Seen from the River, William C. H. Waddell, 1870. (Courtesy New York City Municipal Archives).

atmospheric condition led to illness, "pernicious gasses spread over the island," leaving death in their wake.[19]

Then, at mid-century, people came to believe that a poison, capable of being transmitted from person to person, caused such illnesses as cholera. Though a specific germ would not be isolated for another decade, an outbreak of cholera had been avoided in 1866. The newly formed Metropolitan Board of Health had conducted a successful educational campaign and had enforced the clean-up of streets, cisterns, and privies. Specific deadly diseases were now considered preventable and avoidable—if certain sanitary conditions were met.[20]

As the piers formed walls for still-water cribs that rapidly

68

filled with all sorts of offensive material, preventive measures could conceivably curtail the outbreaks of typhoid and yellow fever in waterfront communities. One way to accomplish this was to change the pier supports to permit currents to wash debris away from sewer outlets. Existing wharves were comprised of a series of wooden bridges resting on large, log blocks filled with stones. In many cases these boxes (or cribs) were placed side by side to form a solid rectangle when attached to the bulkhead. This in effect created a barrier to the washing action of the waterway. At the 1870 Dock Department hearings "open pilings" were recommended instead. A series of tall, supporting columns would be anchored several feet apart, well into the river bottom, and topped by a stone or wood surface to form a finger pier. Under this system river currents could easily pick up the sewer debris and, pushing it between the pilings, carry it out to sea.

A layman, A. D. Bishop, proposed another scheme for handling New York's waste. He connected disease to the acute pollution in waterfront slips and to the dirt that accumulated in the densely packed slums. His solution was to place reservoirs along the bulkhead wall in which the city's sewage would accumulate. The solid matter would be removed, leaving wells of salt water that could be used to wash the streets of "the dense communities."[21]

Novel ecological schemes were proposed to dispose of the solid matter. Receiving basins were designed to be placed under sewer outlets. The manure that collected here could be transported by scow to an open dump as was the practice with dredged material. Alternatively, manure could be sold at a cut rate to farmers who, it was suggested, might pass on the savings to the poor in the form of lower produce prices. The proceeds from sales of sewer contents, it was also proposed, might help pay for the city's waterfront renova-

Figure 5. Design for Docks and Warehouses, New York, View from the River, O. E. Pray. Engineer, 1870. (Courtesy New York City Municipal Archives).

tions. Yet another innovative idea came from a lawyer, John A. Bryan, who recommended piping the sludge to New Jersey or Long Island for landfill. Where would such a pipe be placed? "...on the bridge to Brooklyn now under construction."[22]

Along with contemporary housing reformers, the engineers and seamen who proposed waterfront changes also pressed for improved living conditions in New York. The rapid influx of poverty-stricken immigrants in the mid-1800s had placed a tremendous burden on the city's housing stock. In sections, primarily below 14th Street, families and their boarders crowded into run-down houses vacated by more affluent New Yorkers on their move uptown. People were also packed into equally compressed quarters in newer four– or five-story tenements. Even dank basements, leaky garrets, shops, outhouses, and stables were used for shelter.

A state legislative committee formed in 1856 had investigated the conditions of Manhattan and Brooklyn tenement houses. They found people packed into tiny, airless spaces. Many of them were sick, and drunkenness, uncleanliness, and crime were the norm. The ensuing legislation had not recommended alternative living accommodations, but had focused instead on improving the standards of living in existing tenement buildings and on defining municipal powers of inspection.[23]

The redesign of the city's waterfront opened another possibility for relieving both the housing problem and the thievery and insobriety that spilled onto, and interfered with, the commercial waterfront—slum removal. According to the plan of J. Burrows Hyde, a prominent city surveyor, railroad tracks could be placed on the roofs of new riverfront warehouses to run "around the island . . . transporting passengers as well as goods." The crowds of the slums "would be brought into immediate connection with the outer districts and the upper portions of the city, and thus removed from these crowded tenement houses."[24]

An additional nineteenth-century approach to relieving problems of congestion and illness within the slum communities was to create public areas to which residents could escape for additional breathing space. In 1856 and 1868 the state legislature had set apart land for the city's first large-scale recreation compounds, Central and Riverside parks.

Yet the concept of small, scattered open spaces at the water's edge was as alien to late-nineteenth-century maritime proponents as it had been to the Randall Commission three-quarters of a century earlier. A few enterprising souls leased recreational boats on the East River between 20th and 23rd streets, and private boat clubs could be found in the West 80s. However, in general, where the waterfront was not used for commerce, so-called "nuisance" activities were the norm. Manure yards, slaughterhouses, fat-melting, bone-boiling, and hide-curing establishments crowded between 34th and 46th streets on both rivers. In these districts, on the Lower East Side, and in Chelsea between West 12th and West 23rd streets, social facilities for the neighboring tenements were limited mainly to saloons, churches, and schools.[25]

Ralph Aston, a career naval officer, was the only spokesman in the 1870 Dock Board hearings for planned, waterfront recreation spaces. His brief showed a sea wall that encircled the island. (Figure 6) At mile-long intervals, large granite piers extended into the water, creating opportunities for pub-

Figure 6. Map of Waterfront Improvement of New York City, Ralph Aston USN, 1870. (Courtesy New York City Municipal Archives).

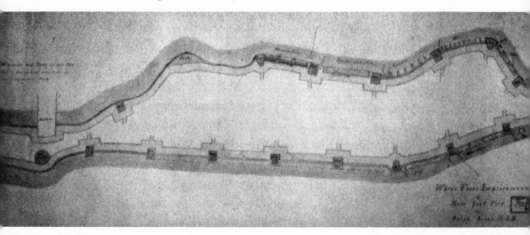

lic squares, promenades, and landings for specific types of commerce.[26] The struggle for waterfront open space envisioned in the Aston plan would be played out in ensuing decades at Riverside Park and over floating pools and recreation piers, for Aston's plan was contrary to the charge of the City Charter that the entire waterfront be used for commerce.

The call for physical growth appropriate to a major maritime city was an overall theme of these hearings. To create an image of civic prosperity in accord with the reality, there was a constant entreaty to push the edges of the island both seaward and skyward. Ignoring the restrictions set down by the State Harbor Commission, proponents of waterfront improvement intended to create structures of solidity and grandeur worthy of a preeminent port.

J. Burrows Hyde's plan for the New York Pier and Warehouse Company was the most elaborate of these schemes. (Figures 7, 8) To create the grandest waterfront in the world, he asked that the Dock Board allow the private Pier and Warehouse Company to usurp sufficient space along the North River, beginning at the Battery, to build a string of twenty piers connected to the land by a new, wider stone bulkhead. The breakwater would straighten the ragged edges of the island, and in places add up to one thousand feet, or five blocks, to its width. On each new pier, Hyde projected a six-story, guarded, fireproof, iron warehouse large enough to house the contents of ten stores. Railroad and elevated tracks would be constructed parallel to the bulkhead on the adjacent outer street, thickening the ring of structures at the city's edge.

A proponent of Hyde's plan described the original version as it was presented to the state legislature in 1869: " . . . it will result in girdling New York with a long line of imposing palaces for the warehousing of merchandise in place of the

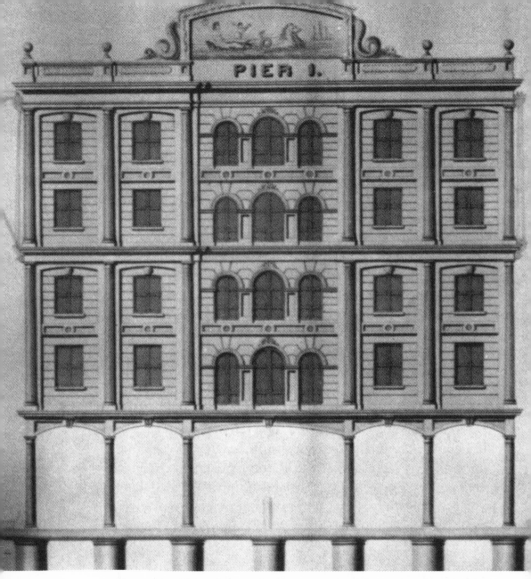

Figure 7. Waterfront Elevation of Pier and Warehouse for the New York Pier and Warehouse Company, J. Burrows Hyde and J. Heuvelman, Architects, 1870. (Courtesy New York City Municipal Archives).

74

rickety, filthy, pestilence-breeding and inefficient contrivances now in use. . . . "[27] This proposal for finger piers, iron buildings, and circumferential elevated structures presaged twentieth-century waterfront development, when a wall was created between the city and its rivers.

The Blueprint, 1871. Less than a year after the Dock Commissioners received the first suggestions for renewal of New York's precious legacies, its harbor and shoreline, an improvement plan was approved; it showed only minor resemblance to the schemes that had been presented. In fact, the official city document, published on 22 April 1871 and known as the McClellan Plan, was just a map showing the bare outlines of a bulkhead and piers accompanied by the engineer's written report. A solid, cement block retaining wall, faced in granite, descending twenty feet below the high-water line and rising six feet above, was to be constructed. The wall would form a semicircular enclosure around the lower half of the island from West 61st Street on the Hudson River to East 51st Street on the East River. This immense girdle would lie outside of the existing bulkhead to allow the widening of the exterior streets. On its water side, sixty- to one-hundred-foot-wide finger piers with "substantial sheds whenever needed" would be constructed at right angles to the island "as the requirements of commerce demand." To eliminate the problem of dirt-infested slips, sewer outlets would be relocated "at or near" the ends of the piers.[28]

Although engineering details were not announced, several stone and concrete docks were envisioned (and built). Pier Number 1 at the Battery, for example, was designed as a series of semicircular arches, faced in granite and decked in

Figure 8. End Elevation Facing Street of Pier and Warehouse for the New York Pier and Warehouse Company, J. Burrows Hyde and J. Heuvelman, Architects, 1870. (Courtesy New York City Municipal Archives).

concrete, leading out into the bay from the bulkhead wall. The arches rested on cement blocks, which in turn were supported by masses of concrete retained in wooden cribs. All of this ultimately rested on bedrock at a depth of some thirty feet in the Upper Bay.

The nature of the bulkhead wall construction depended somewhat on the condition of the river bottom. Yet the general pattern devised by engineer McClellan would be to

76

drive several lines of piles into the riverbed and to dump broken stones and concrete between the piles to form a solid, level, underwater platform parallel to the shore. Rows of huge, six-foot-high cement blocks would be piled on this platform until the top of the bulkhead wall reached six feet above water. Upon completion of the wall, the area between the land and the new bulkhead would be filled in with stone and earth.[29]

The Dock Department's *First Annual Report* discussed why certain of the major concerns among participants in the public hearings were not included in the McClellan Plan. The provision of adequate waterfront storage facilities was referred back to private warehousemen on the assumption that they would "find it most convenient and economical to locate their buildings along the line of the river streets." The waterfront railway was postponed because it could "easily be built" after completion of the river wall.[30]

Unspoken was a tradition of private development that the city would find hard to challenge. In the past, sheds had been built by those leasing docks and bulkheads to conform to individual needs. The Hudson River Railroad had also increased its private presence along the waterfront. It had merged with the New York Central and had added steam locomotion below West 30th Street and a large, new passenger and freight terminal at St. John's Park, near Canal and Hudson streets. With the opening of Grand Central Station for passenger traffic in 1871, there would be even more opportunities to expand its freight connections with pier-ended West Side streets.[31]

Ultimately, failure of the final waterfront plan to be influenced by the public meetings may be attributed to the attitude of the Dock Department leadership. According to the minutes, Chief-Engineer McClellan did not attend a single hearing and dismissed the proposals as technically unfeasible. "The want of knowledge of the ground," wrote McClellan, "deprives them of practical value, except as suggestions for a general system." The Dock Board also commented on the lack of professional expertise to be found in the proposals presented by the public. "The real difficulty," they wrote, "consists in carrying out these ideas practically. This can be done only with the accurate knowledge of the varying conditions of the problem through the long continued efforts of a corps organized for the special purpose."[32] As had been the pattern in European port cities, only the engineers had this special expertise.

The waterfront improvement plan of 1871 did have practical effects. It provided a pattern for the future placement of first-class commercial space as landings could now be increased by 25 percent and the available pier area doubled. Even the hearings were useful. They provided one more platform for expression of the social concerns that were increasingly on the minds of contemporary reformers. The fact too that so many individuals bothered to present their schemes was in itself support for waterfront improvements, and the commissioners admitted that certain ideas did provide additional weight to their conclusions. The hearings also created the illusion that the public participated in the process of planning. It is of particular interest for planning history that in 1871 the public was visionary and the professionals' notions remained within the realm of practicality.

NOTES

1. *Real Estate Record and Builders' Guide*, 16 April 1870, 3; and City of New York, Department of Docks, *Annual Report for 1871*, 35. (Hereafter abbreviated *RRBG*.)

2. *RRBG,* 16 April 1870, 3.

3. City of New York, Department of Docks, Letters, New York City Municipal Archives.

4. The Commissioners of the Sinking Fund of the City of New York, *The Wharves, Piers and Slips Belonging to the Corporation of the City of New York, 1868,* 2 vols. (New York: New York Printing Company, 1868); and *The New York Times,* 10 February 1867, 8, 27 October 1867, 4, and 31 January 1869, 4. (Hereafter abbreviated *NYT.*)

5. Moehring, *Public Works and the Patterns of Urban Real Estate Growth in Manhattan 1835–1894,* 18–22.

6. *Public Meetings of the Department of Docks to Hear Persons Interested in Improving the Waterfront, June and July 1870* (New York: New York Printing Company, 1870), 22. (Hereafter abbreviated Docks, *Meetings.*)

7. N.Y.S., *N.Y. Harbor Commission,* 251–52; and DeVoe, *Market Book,* 447. Dredging costs ranged from $1,700 to $5,000, and would recur within one to three years.

8. N.Y.S., *N.Y. Harbor Commission,* 232; and Docks, *Meetings,* 8.

9. Docks, *Meetings,* 27, 10; and Moehring, *Public Works,* 133–38.

10. New York City, Board of Aldermen, *Documents of the Board of Aldermen of the City of New York,* 1855–70.

11. *New York Times,* 26 September 1867, 2; and New York Pier and Warehouse Co., *Piers and Wharves of New York* (New York: Evening Post Steam Presses, 1869), 46.

12. *NYT,* 16 February 1869, 2.

13. *New York City Charter,* Ch. 383 L. 1870, Article 14; and Section 42.

14. Trows, *City Directory* 1865–75, n.p.; Bonner, *New York The World's Metropolis 1623–4 — 1923–4;* and WPA, *Maritime History,* 31.

15. William H. Harris and Judith Levy, eds., *The New Columbia Encyclopedia* (New York: Columbia University Press, 1975), 1642; and "George McClellan," *Magazine of American History* 14 (1885):242. McClellan served at the Dock Department for three years and in 1878 became Governor of New Jersey.

16. Josef Konvitz, "Engineering and the Spatial Organization of Port Cities, 1780–1830," Paper Presented to a Combined Meeting of the Mid-America Historical Geography Association, The Ontario Geographers, and the Eastern Historical Geography Association, Pittsburgh, Pa., 24 September 1892, 3.

17. Docks, *Meetings.*

18. Ibid., 62–70, 149, 41–48.

19. Pier and Warehouse Co., *Piers and Wharves of N.Y.,* 22; Charles E. Rosenberg, *The Cholera Years: The United States in 1832, 1849 and 1866*

(Chicago: University of Chicago Press, 1952); and Citizens Association, *Report,* xxxvii–xlviii.

20. Docks, *Meetings,* 100.
21. Ibid., 23, 49–50, and 59.
22. Ibid., 23.
23. Robert W. DeForest and Lawrence Veiller, eds., *The Tenement House Problem,* 2 vols. (New York: Macmillan, 1903), Reprint (New York: Arno Press, 1970), 77, 88, 90–92; and Anthony Jackson, *A Place Called Home: A History of Low-Cost Housing in Manhattan* (Cambridge: MIT Press, 1976), 18.
24. Docks, *Meetings,* 23.
25. Citizens Association, *Report,* 120, 186; and DD, Letters.
26. Docks, *Meetings,* 149.
27. Pier and Warehouse Co., *Piers and Wharves of N.Y.,* 9; and Docks, *Meetings,* 20–32, 78–97.
28. DD, *Annual Report 1871,* 40, 38. The placement of the wall was later modified to stop at Grand Street on the East River.
29. Charles K. Graham, *Report for the Year Terminating April 30, 1875* (New York: Martin B. Brown, Printer), 1–9.
30. Docks, *Meetings,* 38.
31. Condit, *The Port of N.Y.: A History of the Rail and Terminal System from the Beginning to Pennsylvania Station,* 39–40.
32. Docks, *Meetings,* 39, 13.

Part Two

Along the avenue which skirts the river, the docks and ships form the teeth of a comb as far as you can see. The arrangement is clear, logical, perfect; nevertheless it is hideous, badly done and incongruous; the eye and the spirit are saddened.

Le Courbusier, *When Cathedrals Were White,* 1936.

CHAPTER 4

Building the Western Wall

Accompanying New York's burgeoning freight traffic was a highly successful passenger trade. By the turn of the century nearly two dozen Manhattan streets ended in ferry terminals. From these piers travelers had the choice of fourteen cross-river routes to connect them with the Pennsylvania, Erie Lakawanna, or one of four other railroad lines that terminated on the New Jersey side of the Hudson. In addition commuters had nineteen ways to reach the newly incorporated outer boroughs. Six private ferry companies, the city, and the federal government provided these crossings at a nominal fee. While railroads had picked up a substantial portion of the coastal and river traffic, piers between Oliver and Market streets on the East River, and at Murray, Warren, Vestry, Canal, and Watts streets on the Hudson continued to be departure and arrival points for travelers to the South, New England, or Albany. Excursions were also popular. In summer passengers boarded boats at East River piers for a day of sun and sea breezes at City Island, Staten Island, or Glen Island, or to fish along the banks of the Atlantic Ocean.

From minute beginnings in 1818, with four minimally out-fitted packet boats regularly carrying travelers between Liverpool and New York, transatlantic crossings by the twentieth century became highly competitive, luxury events. New York in 1899 was described by Idell Zeisloft in *The New Metropolis* as "the focal point to which routes of travel converge and whence they radiate to all important points of destination in both hemispheres."[1] Zeisloft counted 150,000 American cabin passengers (or the entire turn-of-the-century population of Queens) who annually arrived or departed New York. He inventoried over forty steamship lines carrying residents, tourists, businessmen, students, and fortune seekers. These vessels, growing ever longer, heavier, speedier, and more sumptuous, were in addition symbols of national prestige.[2]

Despite the grandeur and size of New York's waterborne tourism, the facilities serving this trade were intolerable. "If the transport facilities for reaching this" port, wrote Zeisloft,

. . . are as convenient, expeditious, safe and comfortable as human ingenuity has yet devised, the same cannot be said of the terminal, landing, and transit facilities for which our city is responsible. The stranger arriving in our magnificent harbor catches a glimpse of tall buildings that impress him with the cosmopolitan bigness and progressive modernism of New York, but give him a poor opinion of our sense of art and beauty . . . he steps out upon the shabby old wharf, encounters rude and venal [customs officers] . . . and cab drivers whose vehicles and manners and methods of business belong to a more . . . primitive age, and looks out on a waterfront as squalid and dirty and ill smelling as that of any Oriental port.[3]

The West Side Building Block, 1880. Planning, which by definition is future-oriented, in practice is often reactive; the Chelsea-Gansevoort plan of 1880 was in response to the ugly setting and long queues, on land and in the river, that Chief-Engineer McClellan had proposed to remove a decade before. In the interim private owners had sporadically improved their waterfronts; however, the new Dock Department had little to show. The East River was untouched, and only three piers and less than a mile of bulkhead had been installed in the Hudson. To add to the problem there were actually fewer piers than in 1870, as the new sea wall had necessitated the removal of several structures in the line of its construction.[4] With these conditions, and with no relief in sight, steamship companies continued to vie for space at the longer piers below Perry Street, where they complained of high rents.

To the north, along the Gansevoort and Chelsea riverfronts, a few crude piers accommodated the many activities that had filled in the blocks between Sixth Avenue and the Hudson River. As the city moved northward toward 59th Street in the late nineteenth century, better residences and commercial establishments filled in the blocks between Lexington and Sixth avenues. Frame houses and four-story brick tenements shared the peripheral blocks toward both rivers with industry. Here shipping services and warehouses, railroad yards, gasworks, and meat-packing establishments were joined by factories attending to the city's domestic life. Ironworks built boilers for homes (and steamships as well); lumber and stone yards supplied building materials; and furniture, lamp, and piano factories, and ceramic and varnish companies provided the amenities.[5]

The piers between West 11th and 23rd streets were intentionally undeveloped, as decades of filling had caused the land here to nestle against the state pierhead lines. Their meager size and sporadic placement not only meant insufficient accommodations for the lesser marine activities in the city, but also created a mile-long barrier between the established shipping center that began at the Battery and the rest of the Hudson River waterfront. Above West 23rd Street were miles of shoreline where with sufficient interest longer piers could easily be built. But the interest was simply not there. Steamship lines elected to crowd downtown rather than to jump past the Chelsea-Gansevoort section where there was plenty of room.[6]

Local businessmen, tugboat owners and trade associations periodically begged the Dock Department to find other remedies, but in reality they hoped merely to reshuffle already taxed facilities. In 1877 a bill requiring more public landings below 14th Street was passed by the State Assembly but vetoed by the governor because the Dock Department said it could "adjust the problem" without legislation.[7]

The adjustment came three years later in the form of a major public works proposal designed by Chief-Engineer George S. Greene, Jr., whose father, a Civil War general, managed the construction of the Croton Aqueduct and was chief of the Topographical Bureau at the Department of Parks. In 1880 George Jr. proposed to augment the McClellan docks project of 1871 by adding piers where none were scheduled to be constructed. Greene's proposal was in all respects but one a straightforward and unadorned dock project: twenty-one piers, eighty feet wide and up to 530 feet long, were to be built on or near street-ends between West 11th and West 23rd streets. Here would be accommodations to suit the *Servia*, the Cunard Line's newest 515-foot-long passenger gem. In addition, rentals would annually add half a million dollars to city coffers.[8]

This was not the first proposal for improvement of the area, for prior governmental bodies had similar perceptions: land here was profitable and water cheap. This was also a perfect spot for a dock system that could bring New York world renown. In 1835 the Committee on Wharves and Piers of the Board of Aldermen had called for construction of a giant wet basin to extend outside of the western edge of the city from Cedar to West 34th Street. Envisioned were three giant stone piers, each a block wide and from a half-mile to a mile-and-a-half long, constructed four blocks out in the Hudson parallel to the shore. Causeways, spaced at intervals to form basins for different activities, would connect the piers to the shore and in addition create protection from currents, winds, and ice. Boats would enter the basins through drawbridges in the center of each causeway.

In recommending this formidable public work, the city fathers had one caveat: the necessary port expansion should not be at the expense of valuable city land. "Should the idea of excavations of any of our Island for wet docks be entertained," warned the committee, "it should be abandoned.... There is no land to be spared... every foot of the island will soon be required for residences, squares, streets, reservoirs, and other uses."[9]

Twenty years later another wet basin was proposed from West 12th to Gansevoort Street. Harbor Commissioners appointed by the state to set pier and bulkhead limits were no longer concerned with excavation but rather with the imminent creation of new land between the existing high-water line at West Street and a proposed thoroughfare that was then below the Hudson, Thirteenth Avenue. Reserve the Gansevoort blocks for wet basins, they counseled, "because it costs little to convert them now while, if they be not so reserved, the increase in property value may induce owners to fill them with earth." If owners filled their water lots, they warned, "the opportunity will be lost of providing additional dock accommodations for this part of the city."[10] It was, in fact, already too late. The section between West 12th and Gansevoort streets was partially filled with the remains of prominent family estates.

George Greene, faced in 1880 with the results of nearly a half-century of municipal inertia and the unwillingness of shipping companies to venture into new, uptown territories, had an unusual proposal. He recommended that all of the filling that had taken place at Chelsea-Gansevoort since the 1830s be undone. To construct new piers that met the requirements of contemporary steamships, part or all of the buildings on twenty-three city blocks would be razed and the blocks themselves excavated and turned back into deep water. (Map 5)

Although the dock administration was free to shape grand visions, it was in fact reliant on passage of three pieces of legislation before the Chelsea-Gansevoort improvements could actually begin. First, the plan of 1871 had to be amended to add piers between West 11th and West 19th streets and to move the bulkhead one block to the east. The McClellan scheme had frequently been revised to allow longer piers or to modify their configuration if the bottom was excessively deep, rocky, or, as in most cases, muddy.[11] A major change of the Chelsea-Gansevoort proportions had not ever been proposed; this would require protracted negotiations.

Next, the state legislature had to decide who would do the improvements. Theoretically, where there was an approved plan the Dock Board had "sole" authority to renovate the

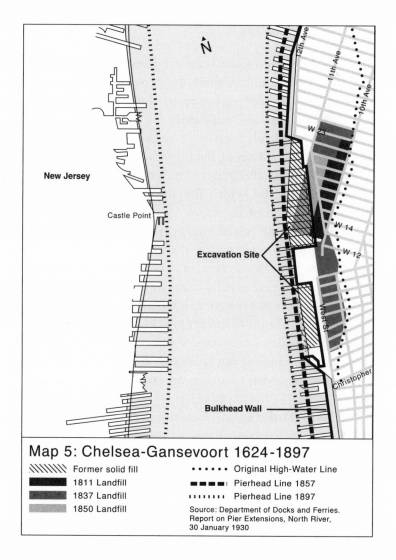

Map 5: Chelsea-Gansevoort 1624-1897

\\\\\\ Former solid fill		•••••• Original High-Water Line	
■ 1811 Landfill		■ ■ ■ ■ Pierhead Line 1857	
■ 1837 Landfill		׀׀׀׀׀׀׀׀ Pierhead Line 1897	
■ 1850 Landfill		Source: Department of Docks and Ferries. Report on Pier Extensions, North River, 30 January 1930	

Labels within the map: New Jersey, Castle Point, Excavation Site, Bulkhead Wall, 12th Ave, 11th Ave, 10th Ave, W 23, W 14, W 12, West St, Christopher, N

waterfront. It constructed the bulkheads and piers and still encouraged owners or lessees to build the pier sheds. Yet the city had given away so much of its riversides over the years that its choice now was either to repurchase all of the Chelsea-Gansevoort section in private ownership, through condemnation—a lengthy and expensive procedure—or to encourage

89

others to do some of the work. To reduce the number of legal proceedings and to share the expense of the improvement, the city needed authority to allow individual proprietors to build piers and bulkheads.

The most difficult legislative problem the Dock Board faced was an extension of the pierhead lines farther into the Hudson, for this challenged the city's commercial supremacy, the very premise on which the new docks were proposed. By the excavation of the blocks from West Street to Thirteenth Avenue, half of the improvements from West 11th to West 13th streets were within the most recent pierhead line approved by the state in 1871. Above West 13th Street in the Chelsea section, even with excavations, the new 530-foot piers would be 170-foot encroachments into the waterway. There was little point in shortening the piers when they were already only fifteen feet longer than the new transatlantic steamers then coming out of drydock. To excavate further to the east in order to make up the difference would be costly.[12]

The New York Harbor Commissioners had set the first boundary lines for New York and New Jersey in 1857 for a specific reason: in this area, the Hudson River was very narrow. Castle Point, a natural, rocky bluff that had been augmented by man, projected into the river from New Jersey. In Manhattan, a half-century of "artificial encroachment" had further constricted the waterway. The distance between West 13th Street and Castle Point, once four thousand feet, had shrunk to three thousand feet. (Map 6) The Harbor Commissioners could have been afraid that the Hudson would eventually dam up and become a lake; however, they voiced their concern over the larger effects of the currents that were obstructed by this neck in the river. These currents were

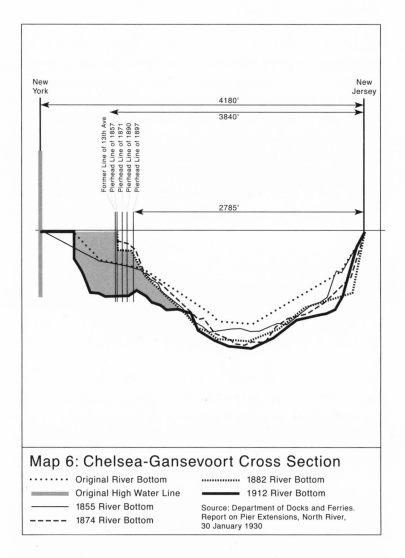

Map 6: Chelsea-Gansevoort Cross Section

⋯⋯ Original River Bottom	▪▪▪▪▪ 1882 River Bottom
▓▓▓ Original High Water Line	▬▬▬ 1912 River Bottom
——— 1855 River Bottom	Source: Department of Docks and Ferries.
⎯ ⎯ 1874 River Bottom	Report on Pier Extensions, North River, 30 January 1930

Labels in figure: New York; New Jersey; 4180'; 3840'; 2785'; Former Line of 13th Ave; Pierhead Line of 1857; Pierhead Line of 1871; Pierhead Line of 1890; Pierhead Line of 1897

deflecting to the west, gradually creating deeper water—and thus a maritime advantage—on the New Jersey side of the Hudson. Instead of washing away Manhattan's shoreside debris, the flow was held back, causing shoals to collect along the southwestern edge of the city.[13]

Laying the Cornerstone, 1890–1902. For ten years the Chelsea-Gansevoort scheme remained inert. Yet the delay proved helpful; with time, people, images, and property values began to change. Several trios of Dock Commissioners had served their terms before any legislation was enacted. In the interim, there was a subtle shift in the commissioners' private backgrounds. In 1880 these so-called nonpartisan guardians of the public good were, in actuality, mayoral appointees with vested interests in dock improvement. Henry F. Dimock, for example, was an agent for coastal steamship companies, while other commissioners had interests in freight-forwarding, law firms that might represent riparian owners, and contracting companies often involved in excavations or providing fill behind new bulkheads. With time granted by delays the men became more political, and thus conceivably had better access to top-level officials who had to approve the Dock Board plans. Commissioner Joseph Koch (1885) was not only a former Judge of the District Court, but a recent member of the State Senate. By 1890 the Dock Board had a strong allegiance with Tammany Hall.[14]

Long considered degenerate, congested, rancid, and immoral, the riversides became the target of both real estate critics and social reformers. In February 1891 the *Real Estate Record and Guide* began an exposé on "our abominable waterfront." "A foreigner landing in this country for the first time," it asserted, "could easily conceive that he had fallen into a semi-civilized community . . . he would see buildings . . . in such a state of dilapidation that they give the impression of a city in its decadence."[15] In this same period the Reverend Charles Parkhurst, President of the Society for the Prevention of Crime, began to publicize his investigation into the brothels that were an adjunct of the maritime community; reporters Charles Farnham and Jacob Riis, investigating conditions on the waterfront and among the poor, wrote of the street Arabs and the homeless who slept in hay barges, sewer outlets, and in a network of crude, subterranean spaces.[16] Redevelopment might remove these undesirables.

Potential opposition by local factory workers to acquisition of waterfront land for shipping purposes had also become diffuse. Gasworks occupied a large portion of eight blocks scheduled for excavation, and their employees would probably not bother to influence their legislators either way. The notorious Tammany political boss, George Washington Plunkitt, made famous by William Riordan, described the problem with this industry: "not very long ago, each gas house was good for a couple of hundred votes. All the men employed in them were Irishmen and Germans who lived in the district. Now, it is all different. The men are dagoes who live across in Jersey and take no interest in the district."[17]

Local property owners too became more amenable to the change, for while valuations and thus taxes continued to rise on properties in the area, returns were low. In fact, the *Real Estate Record and Guide* described waterfront rents in general as lower than in any built-up section of the city. Some owners defaulted on their mortgages to the banks or insurance companies. Others, for example, William Astor, who like his father, John Jacob, had accumulated substantial real estate holdings to the east of Tenth Avenue, offered to sell their waterfront blocks to the Dock Department.[18] These owners looked to the dock improvements to raise the value of their upland parcels.

With time even the government overseers changed, initially bringing with them more liberal policies. By 1890 the United States Army Corps of Engineers, under the War Department, had superseded New York State as arbiter of borders in navigable waters. This added yet another layer of authority to the development process, but now the War Department actually expedited this process by setting new western limits between West 11th and 14th streets. The Dock Department thus acquired from forty-five to 160 additional feet out into the Hudson; and George Greene, Jr., changing his blueprints to eke out more water footage toward West Street, was able to redesign six piers, increasing their length from the old maximum of 530 feet to a new maximum of 728 feet. Ironically, with the 601-foot *Luciana* and *Campania* on their way, the new confines now allowed more than enough space in Manhattan for the largest ocean steamships ever made. However, the city had to retrieve the shipping companies that were moving to new, spacious docks at Castle Point in Hoboken, New Jersey.[19]

In the early 1890s delays began anew, this time in the state legislature. A bill to grant the city the authority to make changes at Chelsea-Gansevoort in the original no-action plan of 1871 failed in Albany in 1888 and again in 1889. Engineer Greene fought valiantly to keep his vision on track. Repeatedly he begged for renewed efforts to obtain legislation to begin the improvements. A new map was finally allowed in 1892, and a year later authority was given riparian owners to develop their waterfronts under Dock Department guidelines and supervision.[20] Now there were two ways to proceed, yet by the beginning of 1894 work still had not begun.

Once the legal obstacles had been overcome, funding became the issue. The Chelsea-Gansevoort improvement involved removing old foundations, dredging the river bottom with a traditional clam-shell bucket or a newer pneumatic pump, removing the mud by scow or pumping it directly onto the shore, and building new piers and bulkheads. In 1880 these were estimated by Engineer Greene to cost

$1,269,000. In addition, as the new docks were to be placed inland on developed, private property, the purchase of 450 lots with buildings and leases added another $3,100,000. Demolition and excavation of these lots might reach $1,055,213. By 1890 the value of city land had risen, causing a 20 percent escalation in Greene's original five-million-dollar calculation.[21]

Approval of the Chelsea-Gansevoort plan was now up to the city. When the Sinking Fund members met in March of 1894 they had to waive the annual three-million-dollar bond limit, which in 1870 had been considered ample for financing Dock Department land acquisition and construction programs. Twenty years later the Chelsea-Gansevoort improvement was estimated to cost double that.

A larger problem was the city's borrowing power. In the early 1890s docks were competing with a variety of other needed improvements: new parks, the Harlem River Speedway, bridges over the East and Hudson rivers, and a long awaited rapid transit system. The city's ability to supply funds for these improvements was limited by the amount it could borrow. The decision on which pressing project to fund thus became a political one. Mayor Thomas Gilroy favored docks, while the Comptroller pushed rapid transit. On 4 March 1894, eight months before his Tammany regime would be voted out of office, Mayor Gilroy prevailed.[22] It was a Pyrrhic victory for the Dock Board. The three-million-dollar bond limit remained, and only half the project (the Gansevoort section) could be approved.

In retrospect, a combination of factors probably garnered the municipal votes to allow this engineering scheme to proceed, factors that illuminate the deeper issues underlying waterfront improvements in New York toward the end of the nineteenth century. The most obvious arguments were economics and pride; new, longer Gansevoort docks would help to avert a commercial shift to New Jersey or Brooklyn, or even worse, Philadelphia. A new public works project also promised such political patronage as construction contracts, which helped the Tammany Dock Board to plead their cause. Approval of the plan may have had social benefits as well. The economy was depressed in 1893, and there was talk of bread lines. At the same time, the public had been aroused by the reformers Riis and Parkhurst. The Gansevoort project promised jobs and minor slum removal. It also provided an image of an improved riverside that could bring a better element of society in daily contact with the waterfront.[23]

Work began on the Gansevoort section in 1894, and for the next eight years the configuration of land and water was in upheaval. Property was condemned, buildings demolished, and earth excavated. A writer in the *Real Estate Record and Guide* described the landscape that just a half-century before had, with infinite patience, been created out of water: "between Bloomfield Street and Fourteenth looks like a Christmas cake after a little boy's teeth has scalloped it."[24]

The cornerstone for the Gansevoort section was laid just after Christmas in 1897. Unveiled by former Commissioner Henry J. Dimock, now President of the Metropolitan Steamship Company (a line running between New York and Boston), the ceremonial plaque commemorated George S. Greene, Jr. as a symbol of continuity in a transient department and as the strongest advocate for his project. It is hoped that Greene, who would soon resign from the Department, enjoyed a moment of triumph.[25] The five Gansevoort piers were opened to the Cunard, White Star, and Leyland lines for their passenger trade in 1902. That same year condemnation proceedings began on the Chelsea blocks.

Elongating the Wall, 1900–1910. At the turn of the century, influenced by the Chicago World's Fair and the Buffalo Exposition, artists and merchants were forming improvement associations and submitting plans for the beautification of New York City. For these reformers, the image of New York as a prosperous port and a major entryway needed uplifting.[26] Shipping statistics alone could not suffice to proclaim the importance of waterfront commerce, especially with contrary visual evidence. The physical condition of the nation's port had not substantially changed since the *Real Estate Record and Guide's* exposé a decade before. The waterfront was a hodgepodge of privately built sheds, encumbered bulkheads, congested streets, and sidewalk saloons.[27] Architectural and aesthetic order was needed. Coherent, impressive buildings lining the waterfront could be solid evidence of civic pride and commercial repute. By their physical harmony with each other, these structures could bring the chaotic conditions of the waterfront under control. The Dock Department's Chelsea plan was a project already on the drawing boards that could produce tangible evidence of the efficacy of these beliefs.

Before the superficial appearance of the Chelsea piers could be addressed the more important problem of their pending obsolescence had to be solved. Plans for this new facility showed eight-hundred-foot-long berths. However, the steam turbine engine, because of its relatively small size, had begun to revolutionize the scale of ships and their operating costs. There was stiff competition to complete the longest and fastest luxury liner, with the 790-foot *Lusitania* and the 822-foot *Titanic* already in production. In anticipation of the *Lusitania,* and of the thirteen-hundred-foot piers on the drafting table for Bush Terminal in Brooklyn, the first change in the Chelsea pier length had been made in 1897. The piers of the already revised 1880 Chelsea-Gansevoort plan were lengthened from 728 to eight hundred feet by carving them inland an additional quarter-block. This plan, however, allowed no leeway for rumors of progress.[28] Redesign of the Chelsea section to accommodate the newest craft was the only way Manhattan would be able to keep its dominant position in the transatlantic passenger trade.

New technology had threatened parts of the island's edge before. Robert Fulton inaugurated the decline of Manhattan's early nineteenth-century East River port with the design of his *Clermont*. Here was a vessel long enough to carry the weight of a steam engine, a boiler, passengers and cargo, and still remain afloat.[29] The size of this new craft and its ability to weather wind and currents to dock anywhere in the region, providing the water was sufficiently deep, made East River slips obsolete and reduced the demand for their use. These same problems were now apparent along the Hudson River at Chelsea.

Thus, in order to provide accommodations for the largest transatlantic steamships in the minds of man, the city proposed in 1903 to build one-thousand-foot-long piers at Chelsea. There was considerable pressure to secure the extra two hundred feet by excising the land further to the east of Tenth Avenue, but the Dock Department remained firm in its opposition to this course of action, citing delays and high acquisition costs. Rationalization, the argument that long ago was used to encourage the filling in of the marginal streets, was now used to prevent excavation. Tenth Avenue, it was said, would become a jagged appendage, exacerbating an already impossible West Side traffic situation.[30]

The Dock Board appealed once again to the War Department to extend the pierhead line two hundred feet, or one

Figure 9. Chelsea Piers Under Construction. View North from the Corner of West and 16th Streets, c. 1909. (Courtesy New York City Municipal Archives).

block, westward. In 1904 this application was denied by the newly appointed Secretary, William Howard Taft, on the historic argument that the waterway was already narrowed "as far as is safe and wise for the interests of the harbor."[31] The Dock Department draftsmen eked out an additional seventy-five feet of pier length by narrowing the adjacent outer street. In this same year, the Cunard Line announced its intention to build a new passenger ship one thousand feet long. The piers under construction were already obsolete. (Figure 9)

In planning, the balance between anticipation of the future and coping with very real contemporary pressures often tips toward the present. At Chelsea, any further attempts to accommodate future marine design would have created inaction; while escalating costs, anticipated lucrative rentals, and the preference of passenger ship companies for a Manhattan location all called for immediate implementation of the outdated plan. Early twentieth-century proponents of civic beautification and city planning provided an unusual and serendipitous source of support.

In 1904 Mayor George B. McClellan (the son of the first chief-engineer of the Dock Department) appointed a temporary commission to make the city more convenient and attractive. Its charge was to gather in one place the most practical of the piecemeal proposals for the betterment of the city and to form them into a unified plan. Included were schemes for bridges, parks, widened thoroughfares, and civic centers. For the Chelsea waterfront, the "New York City Improvement Plan" was a well-timed public relations document. Criticizing the patchwork of structures along the waterfront that were the result of development by individual lessees, the commission cited the Chelsea improvement as an example of "a unified design and construction" that would create "harmony and symmetry" and a "waterfront with an architectural appearance worthy of the city."[32]

Although the City Improvement Commission was short-lived and most of its recommendations were shelved, the Chelsea plan survived and was implemented with decorative details unusual for such utilitarian structures. In a unique departure from past practice where lessees had been in charge of constructing enclosures, the Dock Department had received permission in 1895 to design and build the pier sheds at Chelsea. Instead of relying on in-house engineers, the department hired private architects to create the city's first major passenger ship terminal. Along three-quarters of a mile of the western edge of the city, Warren and Wetmore, who would in this same period help to design Grand Central Station, created conformity and order in the form of an unbroken line of reinforced concrete and pink granite buildings. The two-story, unified facade was designed with triangular pediments that hung over the entryways at the ends of east-west streets. In these, and over the first floor windows, heroic cement sculptures were set celebrating the history of trade. To proclaim the city's international preeminence, huge cast iron globes sat at the apex of each pediment.[33] (See Figure 29, Chapter 6)

On 26 February 1910, thirty years in the making and spanning the administrations of two chief-engineers and innumerable dock commissioners, mayors and secretaries of war, the Chelsea docks finally opened. Heralded as "one of the most remarkable waterfronts in the history of municipal improvements" was a pink and white passenger ship terminal with bronze elevators and "superbly fitted waiting rooms." Here travelers could now board the longest and swiftest ships in the world—the *Olympic* and *Mauretania*. According to one reporter, the tawdry surroundings that formerly greeted visitors to New York were replaced by a "waterfront so imposing that the foreign visitor will no longer find false impressions of the great city . . ."[34]

This project had not succumbed to legal torture or funding strictures, nor had it been abandoned because it failed to keep up with the engineering competition for the biggest and brightest liner afloat. Instead, for the next fifteen years, the Chelsea piers accommodated the "queens of the sea," albeit with their sterns jutting out unprotected into the Hudson.

The piers also made a portion of the waterfront more hospitable for a select group of people; they were a showcase for public pride and a working example of early twentieth-century civic enhancement.

NOTES

1. I. Idell Zeisloft, *The New Metropolis* (New York: D. Appleton, 1899), 104. For an excellent inventory of the activities and occupants along the shores of all of the boroughs at this time see Sidney W. Hoag, Jr., "The Dock Department and the New York Docks," *Proceedings of the Municipal Engineers of the City of New York* (New York: The Municipal Engineers, 1906), 72–92.
2. For works on New York as a passenger port see John G. Bunker, *Harbor and Haven* (Woodland Hills, CA: Windsor Publications, 1979); James Morris, *The Great Port: A Passage Through New York* (New York: Harcourt, Brace and World, 1969); and WPA, *Maritime History*.
3. Zeisloft, *The New Metropolis*, 104.
4. City of New York, Department of Docks, *Annual Report 1880*.
5. *Bradley's Water Front Directory* (New York: David L. Bradley, 1881); DD, *Annual Report 1880*, 78–80; New York City, Department of Docks, Letters, File "West 11th–23rd Street 1870–90," New York City Municipal Archives; *New York Times*, 2 August 1896, 8; and E. Robinson, *Atlas of the City of New York, 1881*.
6. DD, *Annual Report 1880*, 14–15.
7. DD, Letters, file "West 14th Street 1870–90."
8. For the various maps of the Chelsea-Gansevoort Improvement see the following DD *Annual Reports: 1880*, 114–16; *1894–5*, 101; and *1899*, 114–15.
9. New York City, Board of Aldermen, *Report of the Committee on Wharves Relative to the Erection of a Great Pier in the North River*, 7 December 1836, 441; and Edwin Ewen, *Plan and Location of the Great Pier for the North River* (New York: Hayward, 1836).
10. New York State, *Report of the New York Harbor Commission of 1856 and 1857*, 27.
11. DD, Letters and *Minutes*, 1872–1880.
12. For listings of transatlantic steamship sizes 1874–1894 see DD, *Annual Report 1894*, 186–89. DD, Correspondence with U.S. War Department, 4 April 1903, in DD, *Minutes, 1903*, 186–87.

13. N.Y.S., *N.Y. Harbor Commission,* 139–40; *NYT,* 12 November 1895, 9; and *RRBG,* 3 August 1889, 1075.

14. Trows, *City Directory,* 1880–1910. In the back of each edition there is a listing of city officials. Commissioner Charles F. Murphy was the brother of John J. Murphy, owner of the New York Contracting and Trucking Co., and an Alderman and Tammany leader, *NYT,* 8 July 1903, 2.

15. *Real Estate Record and Builders' Guide,* 27 February 1891, 189.

16. Charles Parkhurst, *Our Fight With Tammany* (New York: Scribners, 1895); Jacob A. Riis, *How the Other Half Lives, Studies Among the Tenements of New York,* 1890, Reprint (New York: Dover, 1971), 153–54; Charles F. Farnham, "A Day on the Docks," *Scribners Monthly* 18 (May 1879):34; and Richard O'Connor, *Hells Kitchen* (New York: Lippincott, 1958).

17. William Riordan, *Plunkitt of Tammany Hall* (New York: Dutton, 1963), 62.

18. DD, Letters, File "West 12th–13th St. 1881"; New York County, Register's Office, Re-Indexed Conveyances of the Blocks and Lots of the City of New York . . . , 1880–90, Block Nos.: 643, 645–46, 651, 653, 686–87, (hereafter abbreviated Conveyances); J. J. Astor Rent Rolls, Astor Papers, Manuscript Collection, The New-York Historical Society; and *RRBG,* 7 February 1891, 198.

19. DD, *Annual Report, 1894–5,* 100; *RRBG,* 13 July 1889, 983; and *RRBG,* 3 August 1889, 1075.

20. DD, *Annual Report 1893,* 120.

21. DD, *Annual Report 1880,* 116.

22. *NYT,* 31 March 1894, 3; *RRBG,* 7 October 1893, 398–99; and *RRBG,* 5 March 1898, 405.

23. *NYT,* 8 July 1903, 2; *NYT,* 10 July 1903, 14; *RRBG,* 30 December 1893, 831; and *RRBG,* 5 March 1904, 480.

24. *RRBG,* 12 December 1903, 1078.

25. *NYT,* 28 December 1897, 12.

26. F. S. Lamb, "On the Embellishment of New York City Waterfronts," *Public Improvements* 4 (15 December 1899):75; Nelson S. Spencer, "The Battery," *Public Improvements* 6 (15 January 1900):122–23; A. D. Hamlin, "Architecture and Citizenship," *Public Improvements* 6 (16 April 1900):265; and Jon A. Peterson, "The City Beautiful Movement Forgotten Origins and Lost Meanings," in *Introduction to Planning History in the United States,* Donald A. Krueckenberg, ed. (New Brunswick: The Center for Urban Policy Research, 1983), 40–57.

27. *Bradley's Reminiscences of New York Harbor,* 1898 (New York: David L. Bradley, 1898); and G. W. Bromley, *Atlas, 1899, 1900.*

28. *RRBG,* 11 June 1904, 1371.
29. Morrison, *History of New York Shipyards,* 29.
30. DD, *Minutes,* 1903, 186–87; *RRBG,* 19 July 1902, 83; *RRBG,* 12 December 1903, 1078–79; and *RRBG,* 19 December 1903, 1126.
31. DD, Letters, File "War Department, April–June 1903."
32. *The Report to the Honorable George B. McClellan, Mayor....* (New York; Kalkhoff, 14 December 1904), 7–8 and photographs at end.
33. City of New York, Office of Docks and Ferries, Proposals for Bids or Estimates, Binding Contract and Specifications," 1906, Manuscript Collection, Avery Library; and *NYT,* 27 December 1909, 3.
34. Bunker, *Harbor and Haven,* 132; *NYT,* 27 December 1909, 3; and "Harbor and Dock Improvements," *Scientific American* 99 (5 December 1908):408.

CHAPTER 5

Attacking the Wall

When viewed from the water, depending on the season, more than half of the western edge of Manhattan is a mass of green, or autumnal reds and yellows, or wintery browns and greys: along 8½ miles from West 72nd Street to the northern tip of the island at Spuyten Duyvil, stretch Riverside, Fort Washington, and Inwood parks. Unlike the open spaces at the edges of older European waterfront cities, with their straight, formal lines, the three parks that frame Manhattan's Upper West Side are undulating and naturalistic. They bring trees, grass, and tranquility to this active urban center.

The designation of recreation space along the city's rivers did not come easily; the heritage of commercial beginnings remained very strong. Petitioners for the eighteenth-century Montgomerie Charter requested the expansion of Manhattan Island "for the greater Ease and Encouragement of Trade." Then the Randall Commission in 1811 confirmed the exclusion of recreational spots at the water's edge. In defending the designation of "so few vacant spaces ... for the benefit of fresh air and the benefit of health," the Commissioners reasoned that, because of their expanse, the rivers would always be available for "the convenience of commerce" as well as for "health and pleasure."[1]

The Battery was a model for this conclusion. At first a popular riverfront promenade and by 1800 a small green at the southern tip of the island, the Battery was one of the city's original public places. Here New Yorkers and their visitors could fish for striped bass, take an evening stroll in the hope of meeting a beautiful lady, or see "all the outlets of this great port, and . . . all its shipping come in and go out."[2]

In the first half of the nineteenth century, every other attempt to reserve waterfront acreage for public amenities met resistance on the grounds of lost commercial opportunity. (Map 7) In 1835, the Common Council's Committee on Wharves, Piers, and Slips touted the benefits "to health and ornament" that sea air and exercise would provide if the city were to reserve thirty acres at Stuyvesant Cove, between 13th and 18th streets on the East River, for a park. This project was scuttled when petitioners argued that it would be more democratic to create several small inland squares, and that recreation space would be a waste of "a great front [for shipping] on the East River."[3]

In 1851, a new even larger waterfront site, Jones Wood, on the East River between 66th and 75th streets, was designated as the "new park in the upper part of the City." Its major proponent, William Cullen Bryant, described the site and proclaimed, "we should be glad to see part of the shore without [docks and warehouses], one place at least where the tides may be allowed to flow pure and the ancient brim of rocks which border the waters left in its original picturesqueness and beauty."[4]

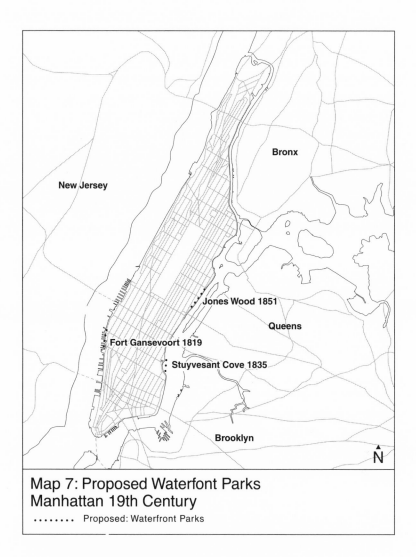

Map 7: Proposed Waterfont Parks
Manhattan 19th Century
.... Proposed: Waterfront Parks

Not everyone shared Bryant's enthusiasm; therefore, the
city appointed a Special Committee on Parks to consider
other locations for its first large-scale public open space. This
group ultimately rejected the East River property because it
was smaller, less accessible, and more costly than an alternate

site at the center of the island. More importantly, Jones Wood was potentially valuable for commercial use, and its designation as a park would deprive the city of increased pier space and maritime jobs and revenue. As the committee noted, "the rapid growth of this city, and its commercial character being its distinctive feature, it would seem to forbid the diminution of its riverfront, which will eventually, and probably very soon, be in demand along this part of the city."[5]

Although waterfront property in general was considered too valuable for public recreational use, there were several small noncommercial inroads at the river's edge. In his *Reminiscences* of early nineteenth-century New York, Charles Haswell mentioned a water-borne theater located on the North River between Spring and Charlton streets. Later, to capture converts in the shipping district, the Seaman's Church sought berths for a floating chapel. Tied to the bulkhead east of Bellevue Hospital was a barge equipped with sundecks. (Figure 10) Here patients on litters were brushed by saltwater breezes that were thought to have curative powers. The St. John's Guild ran another floating hospital, which is still operating today. Their barge took women and children on daily outings to New York Bay; voyagers were given baths, checkups, and lessons on proper health care.[6]

The most incongruous social enterprises were floating swimming pools. As early as 1817, Blunt's *Strangers' Guide to New York* advertised two marine baths located on the West Side near the Battery, and beginning in 1870 "free floating baths" became a municipal institution. Berthed each summer at city piers designated specifically for public use, the pools resembled giant houseboats with open inner courtyards formed of sea water. They offered poor and working-class men and women a supervised place for exercise and pleasure,

Figure 10. Bellevue Hospital Floating Sundeck, 1930. (Courtesy New York City Municipal Archives).

and also provided a facility in which to have an occasional bath.[7] (Figure 11)

By the 1880s the attitude in the city toward reservation of waterfront land for open space had begun to shift. Riverside Park had been designated as park land in 1867, at the request of Upper West Side property owners who wished to be relieved of the expense of grading the steep slopes of the area to the level of the river.[8] On the Lower East Side, where reformers were calling for relief from the rising rates of ill-

Figure 11. Ladies' Day in the Public Bath, East River, c. 1890. (Courtesy Jacob Riis Collection, Museum of the City of New York).

ness, crime, and vice caused by housing congestion, some riverside industrial property was actually vacant. Here, adjacent to some of the city's oldest tenement houses, were the remains of a once-booming mid-nineteenth-century wooden shipbuilding industry; in their heyday, firms had built foreign warships, coastal steamers, and clippers for the China trade. Now, due to the invention of the iron hull, all this had collapsed. Corlears Hook Park was legislated in 1884, and empty property between Corlears, Cherry, Jackson, and South streets was acquired by the Parks Department. However, an unobstructed waterfront setting for the park would not be realized for thirty years; the park legislators had reserved one hundred linear feet at the edge of the East River for commerce, even if it was not vitally needed.[9]

The formal acceptance of a noncommercial activity on the city's wharves began eleven years later with the creation of municipal recreation piers. The landings alongside these specially built piers were still reserved by the Dock Department for boats bringing fresh produce into the city. Meanwhile, the decks, protected by ornamental roofs, offered residents from the tenements a place to play, to receive free milk, and to be taught lessons in American social mores. At night lighted piers on the Lower East Side, in Chelsea, and in Harlem were transformed into band shells and dance halls and were popular entertainment spots for local adults.[10] (Figure 12)

Riverside Park, 1868–1894. The name "Riverside" for the city's first extensive waterfront commons is a misnomer. The park was not immediately next to the Hudson; rather, it sat as a buffer zone between the residential properties on the heights (west of West End Avenue, then Eleventh Avenue) and a commercial transportation strip along the river. The original space, from West 72nd to West 129th streets, was a steep tract of forsaken land. At the park's western edge was a right-of-way that the Hudson River Railroad first occupied in 1847. Beyond this, under the auspices of the New York City Dock Department, whose charge was to promote maritime trade, lay the ragged shore. (Maps 8, 9)

Frederick Law Olmsted proposed a handsome design for this open space. Olmsted's New York reputation as a landscape architect had been earned at Central Park, and his na-

Figure 12. Recreation Pier, East 112th Street and the East River, c. 1899. (Courtesy New York City Municipal Archives).

tional reputation already included designs for waterfront parks and boulevards in Chicago and Buffalo. His new Riverside Park was to be a bucolic green, planted with specimen bushes and trees and bordered on the east by a curved street named Riverside Drive because it overlooked the Hudson.

Olmsted also recognized that the commercial edge would somehow have to be reconciled with the woodland and green. From West 85th to West 89th streets, where Riverside Drive would drop sixty-five feet and brush the railroad tracks, he recommended a plan originated by the Department of Public Works. "Instead of filling up with earth the great space over which the Avenue [Riverside Drive] would need to be constructed, it should be utilized as a building suitable for a market or other public purpose, the walls of which would

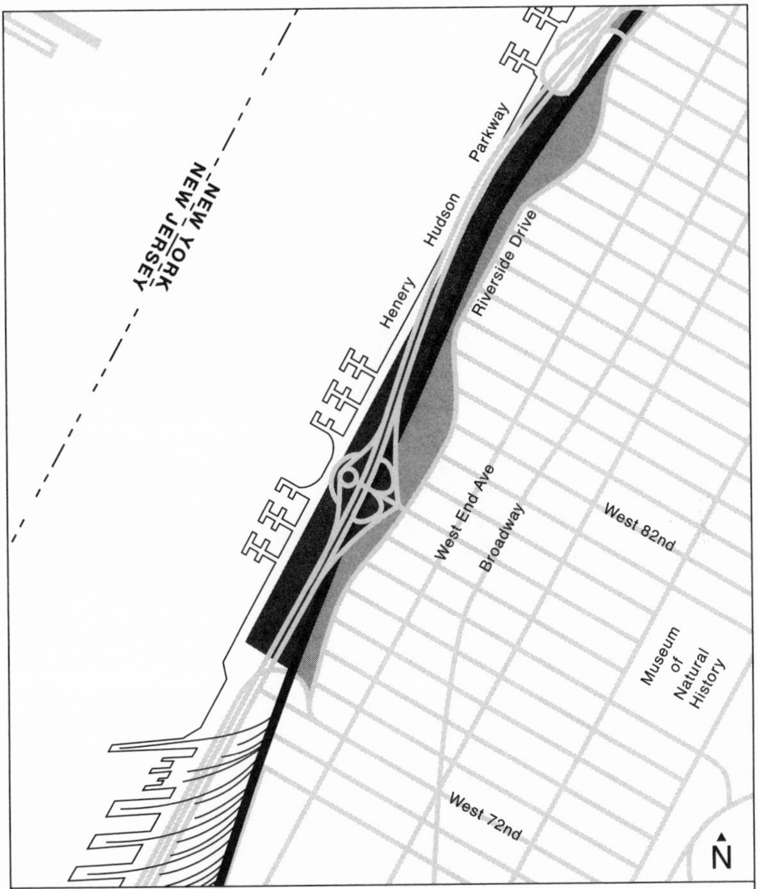
NEW YORK
NEW JERSEY
Henery Hudson Parkway
Riverside Drive
West End Ave
Broadway
West 82nd
Museum of Natural History
West 72nd
N

Map 8: Riverside Park A 1868-1894

Railroad
1868 Park
1894 Park

Source: Frederick Law Olmsted, "Map of the Riverside District," 18 January 1875; and Robert Moses, Memorandum.... 10 June 1935

thus have at this point the character of a terrace commanding a fine view of the river."[11] This element was never implemented. Instead, between 1870 and 1880 shrubs and trees were planted to make the park a screen between the residential neighborhood and the commercial shore. Once complete, the

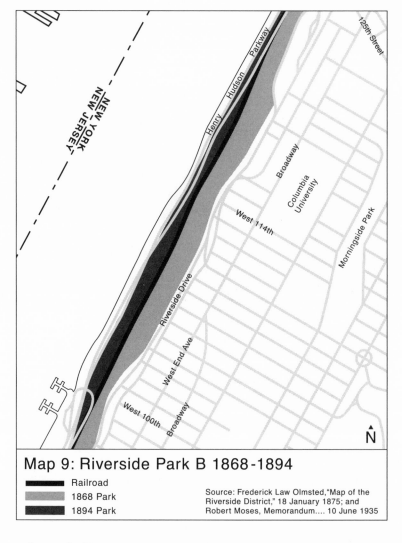

Map 9: Riverside Park B 1868-1894

- ▬ Railroad
- ▬ 1868 Park
- ▬ 1894 Park

Source: Frederick Law Olmsted,"Map of the Riverside District," 18 January 1875; and Robert Moses, Memorandum.... 10 June 1935

N

value of this commons to the neighborhood began to supersede the commercial importance of the undeveloped waterfront.

When the first section of Riverside Park opened in 1880, the adjacent Upper West Side neighborhood had not yet returned to the elite residential district it once had been. Removed geographically from the developed part of the city, which now reached 59th Street, it lacked sewers, electricity, transportation connections, and a wealthy population. Elegant summer homes that once had faced the river, such as the Gerrit Stryker mansion at West 96th Street and the Jacob Mott residence between West 93rd and 94th streets, had burned to the ground. Many of the remaining early nineteenth-century estates had changed ownership several times and by 1880 had become low-income boarding houses.[12]

The park, therefore, was heralded more as a real estate opportunity than as an addition to the city's open space system. "The opening of Riverside Avenue to the public," wrote the *Real Estate Record and Guide,* "has shown how matchless and unique this location is. It brings into the very city all the breadth of prospect, freshness and healthfulness of a summer residence on the banks of the Hudson."[13] The call was for a return of the wealthy elite. It was hoped that the Vanderbilts and the Astors would be diverted here from Fifth Avenue. Yet there was a problem with this premise: numbers. "If the wealthier classes were to occupy the West Side," argued the *Guide,* "it would be a very long time before the region would be settled. There [is] wealth in New York, but as yet not enough to fill the whole West Side with ornamental residences."[14]

The new park effected some improvement, but local real estate failed to stabilize. Between 1879 and 1881, ground was broken for a few "fine houses" on Riverside Drive, but many vacant parcels remained and property values continued to fluctuate. Among the stated reasons were a continuing lack of adequate transportation and the river winds, which were so strong that it was difficult to heat the houses in winter with ordinary furnaces.[15]

In addition, this residential area lacked commercial restrictions; the unfinished waterfront lay perilously close, leaving room for the undesirable aspects of city life to intrude. The new park failed to contain these nuisances. Between 1870 and 1890 the merged New York Central and Hudson River Railroad increased its Riverside holdings, filling in the Hudson River to the west of its tracks from West 72nd to West 76th Street and erecting a dock and freight depot. Above this, to West 79th Street, the Dock Department constructed a bulkhead.[16]

The growth of the Upper West Side, either real or anticipated, caused further commercial incursions. Local contractors vied for the privilege of dumping the excavations from new streets and residences behind the Hudson River bulkhead, thus making more usable land for the Dock Department to offer for rent. In the 1880s this department assigned space at West 79th Street to the Street Cleaning Department as a dump for the debris accumulating from the growth of the city west of Central Park. It even gave landings for building materials at West 79th and 96th streets to the West End Association, founded in 1866 "to promote West Side improvements and protect the interests of property owners west of Central Park."[17]

Several years later when this group petitioned the state legislature to extend Riverside Park to the river, it deplored the rotting piers, dirty lumber, mounds of debris, and old sheds along the waterfront, as well as the elements residents hoped to remove—crooked dwelling houses, shanties, and a saloon.[18] (Figure 13) The petition had been prompted by federal action in 1890, when the United States War Department extended the bulkhead boundary below the Riverside neighborhood from four hundred to five hundred feet into the Hudson River. Just as the locale was beginning to assume elegance, the specter of new land on fill to the west of the

Figure 13. Timber Basin at West 72nd Street, 1903. (Courtesy New York City Municipal Archives).

railroad tracks raised fears of new intrusions: factories, and asphalt and gasworks whose smoke and smell would further depreciate local property. The same association that had earlier requested increased dock space reversed itself and pushed for a truly "riverside" park, all the way to the water's edge. The rationale was not benevolence—that the city needed more open space—but that a waterfront greenbelt was important to the preservation of residential property values.[19]

Spurred by the rumor of a new bulkhead line, local residents and landowners produced a flurry of designs to show how the newly made land could be developed. Some schemes reflected a continued dichotomy between open space and waterfront commerce; others, following Olmsted, tried to

117

reconcile the two. Of the latter, Peter B. Sweeny's *Rotten Row* was an exaggerated preview of the present Riverside Park. Sweeny, twenty years earlier a Park Board member of the infamous Tweed Ring, proposed a bilevel structure parallel to the river, with a lower street for commercial traffic, and an upper-level terrace for pedestrian walks, a bridle path, and an uninterrupted avenue for "fast driving" [of horse-drawn carriages].[20]

When the bounds of Riverside Park were finally extended into the Hudson River in 1894, it appeared that park, and therefore residential, interests had prevailed. (Map 8) Most of the frontage between West 72nd and West 129th streets to the new bulkhead line was transferred to the Park Commission. However, either because the Upper West Side residents were also still reliant on the waterfront for cheap transportation or because they compromised in exchange for legislative and Dock Department approval, two shoreside strips a total of four blocks long, one from West 77th to West 79th Street and the other at West 96th Street, were placed under the jurisdiction of the Dock Department. These reservations would haunt park advocates in the ensuing years.[21]

Who Plans? 1910–1916. The landfill and the accompanying late-nineteenth-century schemes to expand Riverside Park to its designated boundaries in the Hudson never came to fruition, and the green remained barricaded behind the New York Central and Hudson River Railroad tracks for another sixteen years. Then in 1910, because the railroad aggravated traffic conditions downtown, attention was focused once again on this Upper West Side riverfront. The Central was doing extremely well. It still provided the only direct rail freight line into Manhattan, and by 1910 its freight loads were

Figure 14. New York Central Railroad, St. John's Park Freight Station, November 1910. (Courtesy New York City Municipal Archives).

increasing a million tons a year. Its two riverside tracks beginning at Spuyten Duyvil now fanned outward into a multitrack yard at the southern end of Riverside Park, from West 72nd to West 83rd Street. Below the park, from West 60th Street to Saint John's Park, smoke-belching steam engines pulling cars laden with milk, hay, grain, produce, and coal rumbled down the surfaces of Eleventh and Tenth avenues and West Street, threatening the lives of pedestrians, interfering with vehicular traffic, and stunting the economic growth of the Lower West Side. (Figure 14) The combination of the railroad's expanded trade and the continued residential and commercial growth of the city toward West 72nd Street spelled trouble.[22]

The state legislature had tried several times to rectify the so-called "West Side Problem," even by outlawing steam locomotives at grade. But no design for achieving this was put forth until 1911, when the state ordered the New York Central to submit a plan for the elimination of its facilities from the surface of public streets and places within the city line, including the removal of the tracks at grade below West 59th Street. Expansion of the existing railroad yards and tracks in Riverside Park became bargaining points for downtown grade crossing elimination, and the nature of the development above the tracks prompted new designs for a waterside park.[23]

The likelihood of a private redesign of the railroad properties, and availability of a large amount of free rock, spurred the city once again to consider the extension of Riverside Park. In 1910, excavations for the Catskill Aqueduct, designed to increase the city's water supply, reached upper Manhattan. Charles B. Stover, Commissioner of Manhattan Parks, who was in charge of the maintenance and design of Riverside Park, negotiated with the Catskill contractors. In return for a permit to sink shafts in other city properties, the water system gave Stover a free supply of rock for Riverside Park. This new fill, while finally expanding the public area to the river's edge, actually inaugurated a second recreation space, separated from the original Olmsted park by the open railroad tracks and yards. The challenge became the design not only of the fill but of the connection.[24]

The title of the first proposal for organizing the land to be created from the river was misleading. *The Joint Report on Proposed Reclamation of Land between 81st and 129th Street* was produced in consultation with the Parks Department. However, it was a Dock Department engineering document, incorporating maximum expansion of the commercial edge with enlargement of the railroad facilities within the park. The presiding Dock Commissioner, still the landlord of several thousand feet of waterfront reserved in the 1894 parks legislation, wished to increase his territory to a continuous strip along the entire bulkhead line of Riverside Park. Here would be room for more pier facilities and the northward extension of Manhattan's westernmost commercial street, Twelfth Avenue, which then stopped around West 60th Street. The report provided an excellent lesson in retaining wall construction; in addition, it showed how the maritime facilities that were already established along the Lower West Side waterfront were to be augmented and extended to meet twentieth-century shipping techniques. Between the original edge of the park and the newest bulkhead, in places 320 feet to the west, the *Joint Report* advocated eleven railroad tracks, a vehicular roadway, two loading platforms, and a wharf— a plan that made the dreamers of the 1870s seem conservative. In addition, a series of sheds, the upper deck of which was marked "Esplanade for Park Development," would cover this freight and transportation network.[25] (Figure 15)

But not everyone was happy with the newest vision. When the *Joint Report* was published in 1911, the *Real Estate Record and Guide* exclaimed that "a crisis has arisen in the history of Riverside Park." With the subway at last providing a rapid and accessible means of transportation, the Upper West Side had enjoyed a decade of residential building activity. The streets between Broadway and Riverside Drive were paved, and most blocks between West 72nd and 100th Street were filled with brick row houses. While some of these were "extravagant," it was the sprinkling of elegant mansions on Riv-

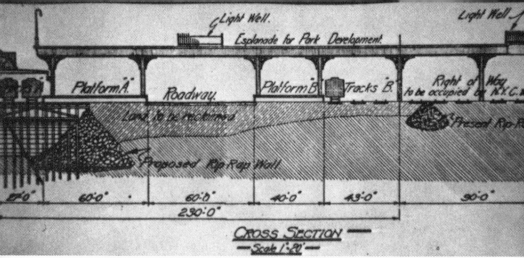

Figure 15. Dock Department Plan, 1911. New York City, Department of Docks and Ferries, *Joint Report on Proposed Reclamation of Land Between 81st and 129th Streets, North River,* No. 5, 27 December 1910. (Courtesy of the author).

erside Drive that raised expectations about the quality of the area. Financier Gustave Schwab's $3 million house, begun in 1902, occupied the entire blockfront from West 73rd to West 74th Street. The estate of the city's Episcopal Bishop, Henry Codman Potter, sat at the corner of West 89th Street. It was hoped that these houses would finally establish this section of the city. Maybe it would never be ultrafashionable, like Fifth Avenue, but it would provide dwellings for those with considerable incomes.[26] (Figure 16)

A new type of dwelling, the twelve-story apartment house, was also introduced into the neighborhood, causing Upper West Side real estate values to rise precipitously. Assessments on the drive and West End Avenue had increased an average of 20 percent in 1910 alone. In one case where apartments were built, the worth assigned to the improvement tripled. (Figure 17) Despite this rapid change, the open railroad tracks and the threat that shipping would bring commercial vehicles into the area created extreme anxiety about the future worth of Riverside properties. Apartment house owners were worried that the new buildings would fail to rent, while neighbors

Figure 16. Riverside Drive Homes, c. 1925. (Copyright ©
collection of The New-York Historical Society).

feared that units would be leased to an undesirable class of
tenants.[27]

To mollify the neighborhood and to meliorate the effects
of dock and railroad expansion on territory that was legally
its own, the Parks Department produced a counter design.
It hired landscape architect A. Van Buren Magonigle to elaborate on the open space that was given short shrift in the
engineering plan. Magonigle had recently designed a water-
gate at 111th Street and the Hudson River to celebrate the
centennial of Robert Fulton's steamboat invention. His
scheme for the Parks Department, published in September
1911, made the transportation-commercial package palatable
to the local community by decorating the dock and railroad
footings, which so long ago had been established in the park,
with attractive open space. There was something in the blue-

123

Figure 17. Riverside Drive Apartments, West 79th Street, 30 May 1915. (Courtesy New York City Municipal Archives).

print for everyone—a recreation pier atop a boat landing and dumping facilities, and athletic fields on the flat roof of the widened railroad tracks. Along the shore, where private yacht clubs and rickety swimming facilities also had historic hold, there were even "opportunities for boating [which] will be greatly appreciated at a time when there is more general interest taken in water sports than ever before."[28]

Magonigle was one of several architects who at the turn of the century tried to deal with Olmsted's pastoral space, the adjoining, noisy, smelly, open railroad tracks, and the still incomplete riverfront property, as one continuous park.

These early landscaping schemes, which grew out of the same city improvement beliefs that influenced the design of the Chelsea piers, perceived the waterfront as a showplace for visitors. Beautification and monumentality were of prime concern. The drawings show sweeping promenades connected to the river and park by grand staircases and formal gardens decorated with heroic statuary and fountains.

The ideas of twentieth-century recreation proponents were also included in these turn-of-the century visions. Riverside Park, which had earlier been viewed as a purely contemplative space in which to relax, was increasingly seen as a place for more active sports, many of which had little relation to the water. In a paper on the "Planning of Cities," Milton See, architect of the Museum of Natural History, described possible bicycle tracks, bathing pools, playgrounds, and open-air theaters. In all of these park plans the railroad was suddenly absent, hidden behind elaborate colonnades or an imaginary embankment.[29] (Figure 18)

But the railroad was in reality still very much in evidence, and between 1911 and 1915, with pauses for litigation and political change, the railroad company and the Terminal Committee of the Board of Estimate negotiated over a series of alternative proposals for disposal of the tracks on both the Lower West Side waterfront and in Riverside Park. That the focus continued to be on commercial development rather than parks was evidenced by the membership of the committee: the Mayor, Comptroller, and the Commissioners of Public Works and Docks.[30]

The annual reports of the dock and park executives were even more revealing of this conflict. Dock Commissioner R. A. C. Smith assumed that his was a primary role. "I am necessarily the technical advisor of this Committee," he wrote, "charged through the departmental force with work-

Figure 18. Scheme for the Embellishment of the City on Its Western Side, by Milton See, 1899. *Harper's Weekly.*

ing out the details of the very important plans for port and terminal development which that Committee is called upon to consider."[31] Charles Stover, on the other hand, bemoaned his lack of direct participation, but the more salient truth was that his department had no funds for this project. "The Parks Commissioner," he wrote,

... is not ex-officio, a member of this Committee despite the fact that there are miles of waterside parks and bulkhead esplanades along the shores of the river and harbor, whose areas must be affected by any general scheme for harbor or port development. It must be clear that no large plans involving waterfront changes should be entered into without there being considered the point of view of the Park Department and its advisors on questions of park development and landscape design.[32]

On 29 April 1916 an engineering plan for the "West Side Improvement" acceptable to both the city and the New York Central Railroad (the newest and final name for this transportation system that had begun as the Hudson River Railroad) was circulated to the public. It was a less glossy version of the 1911 Dock Department Plan. In return for grade eliminations downtown, the railroad would increase its Riverside tracks to six, extending them one hundred feet eastward into the park. In addition, it would enlarge its yards to the southern end of the partially filled, new parkland. In order to modify the effects of these new intrusions into the park, the railroad agreed to erect a steel and concrete box to cover its facilities from West 72nd to West 129th streets, and to underwrite $300,000 worth of landscaping on this structure.[33]

Despite five years of discussions and designs, the waterfront park still remained a vision. As shown the railroad route, while under construction, would devastate acres of specimen trees, and when covered, would create an even wider and higher barrier than already existed between the Olmsted and waterfront parks. The funding allotted by the railroad was only half that estimated for proper restoration and landscaping. Moreover, there was serious question about the ability of the concrete covering (punctuated by thirteen ventilators, each ten feet high) to hold enough earth to maintain anything but grass.

Outshore, where since 1911 mounds of rock had been dumped at random, the situation was equally bad. Long expanses of railroad tracks, occasionally separated from the water by projections of unleveled land, still clung to the river's edge. (Figure 19) In the latest railroad plan, "not an ounce of soil, not a blade of grass, not a single bush or shrub, not a foot of promenade, pathway or steps was to be provided" for this section. Once the railroad work was complete a

Figure 19. Riverside Park Landfill, 1915–1924. (Copyright ©
collection of The New-York Historical Society).

"stretch of broken, barren rock, left ugly and untouched"
save for two large, stucco refuse disposal buildings on Dock
Department property at West 72nd and West 77th streets
would remain.[34]

Who Listens? 1916–1918. A month after the city and the New
York Central announced their agreement public hearings began, airing a plethora of conflicting concerns. The outpouring
was unlike the one-sided experience with earlier Dock Department plans, first in 1870 and later at Chelsea, where proponents of commercial improvements uniformly backed city
efforts to develop the shore. There were those who championed the newest plan for Riverside because it was a part of
larger track changes that would finally clear downtown West
Side streets and, it was believed, improve freight facilities

128

and declining Lower West Side real estate values. For these chiefly trade groups—the Merchants' Association, the United Real Estate Owners Association, and the Board of Trade and Transportation—New York was "primarily a port to which aesthetic conditions [at Riverside Park] must give way."[35]

In the middle stood the arts groups, associations of architects, landscape architects, and the Municipal Art Society, founded during the City Beautiful period. They allied with such reform associations as the City Club and the Citizens Union. For these organizations, the latest agreement was better but not ideal. They supported the removal of the "Death Avenue" tracks, but not at the risk of Riverside Park. They asked for "more time to consider all conditions."[36]

Opponents of the city agreement with the railroad were concerned primarily with preservation of parkland and Upper West Side realty values. Included were local chambers of commerce such as the West End Association and the Washington Heights Taxpayers Association. These groups feared that the expanded railroad yards and commercial streets along the river north of West 129th Street would adversely affect their living conditions.

There was, in addition, one other organization that would turn out to be a formidable opponent, the Women's League for the Protection of Riverside Park. The league was founded on 19 May 1916, to "protect the territory included at the present time in the area of Riverside Park, to keep it forever inviolate for the use of the public." Open to "any woman interested in promoting the objectives of the league" who paid one dollar a year, the group was described by the press as an "association of *prominent* women."[37] Although it is impossible to assess accurately, the early membership appears to have been mixed. A small proportion of their names can be found among lists of "notable" professionals, merchants, etc., of New York. Some of these, the William Rhinelander Stewarts, for example, were descendants of established New York families. Others, such as the Hearsts and Guldens, represented new power and wealth. Philanthropic activities were common to the founding members of the league. Some, for example, championed disadvantaged youth by serving as presidents of an orphan asylum and a home for crippled children and as members of the City Playground League. Many were active in women's clubs. President Nanette Bryan chaired the Parks Committee of the Women's Municipal League, of which vice-president, Helen Culver Kerr (daughter of Andrew R. Culver, a nineteenth-century developer of another waterfront, Coney Island) was also an official.[38]

The early members of the league were also primarily West Siders and thus had a proprietary interest in protecting Riverside Park. Nearly a quarter resided on West End Avenue or Riverside Drive. Several of the women, through marriage, must at least have been aware of neighborhood concerns. Mrs. Henry Everston Cobb was the wife of the Pastor of the West End Collegiate Reform Church, and the husbands of Mrs. Charles L. Craig and league vice-president Mrs. John C. Coleman were leaders of the West End Association. Mrs. Joseph Paterno and Mrs. Abraham S. Post were wives of a local developer and a realtor. For some active members, however, commitment had no residential boundaries. Vice-president Mrs. James Stewart lived in Scarsdale. Helen Kerr, originally a West Sider, devoted nearly two decades of her life to the league from a home on Park Avenue.[39]

Curiously, the members of the league were generally older and had fewer children than would be expected from a group interested in a local park. Among the better known couples, many of the husbands were over sixty. Of the fifteen most active women in the association, three were unmarried and none of the remainder had small children.[40] Age in itself may have been a reflection of the ability to donate time and the funds to underwrite the cause. Yet, together with the lack of young children, it reinforces the view that the park at that time was not regarded primarily as a recreational resource, but rather as a buffer to protect local residences from encroachment by undesirable urban activities.

As it began to pursue its stated objective—the preservation of Riverside Park—the league used arguments that were persuasive but diverse. Some contentions—for example, that the park helped save children's lives during an infantile paralysis epidemic, or that thirteen thousand school-aged children would be deprived of a place to play during the years of railroad construction—were made specifically to attract public health and recreation proponents to their cause. Other approaches, such as a discussion of the threat of lowered property values and the unquantifiable costs of the despoliation of trees and shrubs, appealed to the pocketbook.[41]

However, the initial main focus was on an "aesthetic" argument, and this was elaborated under the banner of the avant-garde women's movement. In the months following its formation, the founders of the league embarked on a massive membership and educational campaign in which they appealed strictly to females. Starting with the established base of the women's clubs, the spectrum was widened with invitations to (and acceptances from) barefoot aesthetic dancers, settlement house leaders, directors of female-run clinics for children, and representatives of women's church groups and artistic salons. By the end of its first year, the Women's League could boast of strength in numbers; 482 members represented a constituency of eighty thousand women who opposed the plans of the railroad in Riverside Park.[42]

Members of this large following proceeded to defend Riv-

erside Park from a distinctive feminist point of view. The
league sought to present the "women's side of the [park]
question," which heretofore had been espoused only by all-
male organizations such as the West End Association. While
suffragettes fought for national political equality on a platform of women as spiritual creatures—distinct from men—
the league, seeking to affect local decision making, imbued
women with a special aesthetic sense through which they
"could see farther than men."[43]

The women were never able to define these beliefs very
clearly. Instead, they turned to Jens Jensen, a prominent Vic-
torian landscape architect and superintendent of Chicago's
West Park District, to help them explain to the public and
to the Board of Estimate the importance of the aesthetics of
Riverside Park to the people of New York. "From the stand-
point of art," wrote Jensen,

> . . . it is a masterpiece, a living out-of-doors art exhibit. . . . It is the
> most precious piece of park land, the foreground to one of the
> greatest views of this country; it affords to the citizens packed away
> in tenement dwellings something of an outlook into the world,
> something of a vision that broadens their horizon and imprints
> upon their souls some of the grandeur of our country. . . . It is in
> our parks the city dweller finds himself; and since it rests with him
> alone to make city life healthier, more beautiful and more worth
> while, then the greater the artistic expression from which he receives
> inspiration, the greater its value to mankind.[44]

However progressive the women's movement might have
been during this period, this statement was a return to nineteenth-century open-space values. It was a conservative appeal for beauty in a time of rapid change when amenities were
considered an unnecessary expense.

Once its membership was in order, leaders of the league

Figure 20. Women's League Protest Poster, c. 1916. (Copyright © collection of The New-York Historical Society).

used political tactics to press for their goals. They held mass meetings; wrote letters and sent flyers to members, public officials, and the press. They assiduously attended hearings, circulated petitions, and with the help of experts conducted inventories of shrubs and trees. They even wrote the National Park Service to inquire if Riverside could be designated a National Park.[45] (Figure 20)

Because of their persistence and success in gathering a following, the league was immediately treated as an adversary by the decision makers. Spokeswomen were harassed, ignored, and intimidated. The first public meeting at which the league could present its point of view was deliberately set in late July of 1916, at the height of an infantile paralysis epidemic. Most middle-class families had fled to the country. When, undaunted, the league did testify at these hearings, members of the Board of Estimate interrupted, whispered, and laughed. The following January when the Ports and Terminals Committee, in charge of negotiations with the railroad, led the press on a tour of the park, it refused to let the women participate. Weeks later, Comptroller Prendergast, chairman of the committee, called the Jensen report "outrageous, untrue and stupid."[46] Still persevering, the women calmly and thoughtfully rebutted each attack and continued their fight to save the park.

By 1917 the battle was no longer in the hands of women alone. A widening spectrum of citywide businessmen's associations, whose major concern was solving the railroad problem downtown, pitched in with the Women's League to counter the testimony of the opposition. Included were the Reform Club, the Institute for Public Service, the Fifth Avenue Association, and the Fine Arts Federation.[47] Then, with the serendipitous support on the Board of Estimate of Manhattan Borough President Marcus M. Marks, the league was able to modify the city's prorailroad position.[48]

Expressing its willingness to change, on 19 March 1917 the Board of Estimate appointed a committee of business representatives, engineers, and technical experts to advise and offer alternative schemes. The women, of course, were not represented. Nonetheless, by the end of the year a compromise plan was passed that met with the league's approval. According to this scheme, the railroad would be allowed to expand its right-of-way westward onto the new land to accommodate two additional tracks. In addition, its entire route through Riverside would be recessed "so as to cause as little disturbance to the park contour as possible" and "roofed over in such a way as to permit the roof to be used either for park purposes" or, anticipating the future, "for a park boulevard."[49]

The railroad refused to approve this plan, and by law the matter was taken out of the hands of the Board of Estimate. In December 1917 the State Public Service Commission was put in charge of the city's West Side railroad problem; the following month the federal government, using its war powers, took over the nation's railroads, moving all proceedings to Washington. Shortly thereafter, as had happened to eighteenth- and nineteenth-century mercantile New York, war curtailed all expansion activities, and the West Side Improvement was temporarily shelved.

In its first two years of activity, the Women's League for the Protection of Riverside Park accomplished more and less than it had expected. The league and associated groups effectively stopped the railroad's plan to intrude into the park, but physically nothing had changed. Children continued to coast down their favorite sledding hills, landing on the railroad tracks. The noisy steam engines continued to carry odoriferous, bellowing cattle on this line, which, where there was landfill, split the park in two. On the waterfront, the heavily used garbage disposal stations continued to punctuate an unfinished wasteland.

Decorating the Wall, 1924–1929. In the early 1920s certain municipal problems were resolved that eased the way for the creation of a "riverside" park. Initial attempts between 1921 and 1923 to legislate the removal of the garbage dumps that had become entrenched at West 77th and West 96th streets were impeded by the Street Cleaning Commissioner, whose office continued to exert a powerful influence on the configuration of the waterfront. This time he asserted that it would cost nearly a hundred thousand dollars extra to hire carts and drivers to carry Upper West Side refuse to dumps outside of the area. By 1923, however, one of the city's first public incinerators was installed on the Harlem River at East 139th Street, allowing the dump at West 77th Street to be closed.

Figure 21. Dump, Riverside Park at 77th Street, 6 June 1929.
(Courtesy New York City Municipal Archives).

Yet the sanitation building, which when constructed in 1915
was promised as "an artistic structure suitable for refuse disposal *so long as it may be needed for the purpose and then convertible
into a shelter . . .* for mothers and children [to] find shelter from
the summer sun" (author's italics), remained "in vacant ug-
liness" as a decaying remainder of a piece of outdated waterfront furniture.[50] (Figure 21)

As sanitation problems found solutions elsewhere, so did
shipping. A new transatlantic terminal was opened in Staten
Island; in 1922 the first of four giant passenger piers, each a
block longer than those at Chelsea, was leased at West 44th

Street to a private shipping company. One thousand feet long and able to accommodate the newest "leviathan," the pier was rented for $270,000, or triple that of a Chelsea pier. The provision of additional pier space, combined with an overall decline in the value of waterborne exports and imports in the Port of New York after World War I, removed the threat of the extension of large maritime facilities into Riverside Park. However, the need for the small public piers for landing coal and supplies that were indigenous to the Riverside waterfront remained.[51]

With progress on other fronts, efforts to resolve the New York Central Railroad problem resumed, and the effects on the park were mixed. In 1920 J. Bleeker Miller, long a champion of Upper West Side real estate interests and now Assistant Corporation Council, instigated a suit on behalf of the city to determine the railroad's rights in Riverside Park. The disappointing outcome of this suit established that, due to its length of occupancy, the New York Central had a perpetual easement. Attempts to curtail the railroad's emission of noise and smoke as it passed through the green met with greater success. As had happened in Chicago with the Illinois Central a few years before, a bill passed the state legislature in 1923 requiring electrification of the New York Central rail lines. Although the time for compliance was periodically extended, the new system would have a significant, positive effect on the environments of the park and the adjacent neighborhood.[52]

There were also administrative changes during this period. In 1921 a new nongovernmental planning and development agency, the Port of New York Authority (Port Authority), comprised of representatives from New York and New Jersey, was established. Matters concerning transportation in the port could now formally be approached from a regional perspective. The Authority, in addition, had financing powers that were independent of city or state debt limits, and this would allow it to implement its schemes quickly. However, existence of the agency would add an additional, often competitive, layer to the administration of Manhattan's waterfront.

After a delay of seven years, planning for Riverside Park was once again revived in 1924, ostensibly because a new landscape design for the landfill and railroad cover was published by the city on January 30th. Stone and money also gave new impetus to the project. The pending electrification of the railroad revived hopes that the New York Central might fund all the necessary park improvements, and once again fill was available. In addition, the Board of Transportation, to effect substantial savings in carting costs while building the Eighth Avenue Subway, asked permission for its contractors to dump rocks along the waterfront, and thus again to extend Riverside Park.[53]

The newest design was also payment by City Comptroller Charles Craig of a political debt to the Riverside community. Although produced by the Parks Department, the plan was sponsored by Craig's office. In an amusing political twist, Craig, a valiant legal defender of the park, had succeeded William Prendergast, chairman of the prewar Ports and Terminals Committee, which had favored railroad expansion. A resident of Riverside Drive and former head of the Law Committee of the West End Association, Craig was also the husband of a member of the Women's League for the Protection of Riverside Park. This organization had lobbied actively twice to secure his election and re-election as Comptroller.[54]

Uncertain Riverside property values also prompted the resumption of park planning in 1924. The previous year, a boom time for real estate and one in which car registration in New York rose 15 percent, Riverside realty had remained at the level of 1910. Between West 69th and 89th streets on Riverside Drive, facing the open railroad yards and the Dock Department piers, values had declined between 2 and 9 percent.[55] As the park had still failed to fulfill its promise to local real estate, a plan for its improvement was a welcome public relations document.

The Craig Plan was the first of four landscaping schemes in the 1920s for a unified Riverside Park. Though none of these was implemented, they show some of the controversial issues that Robert Moses would resolve in the next decade. One aspect of the Craig design was the inclusion, on landfill near the river, of large playgrounds and athletic facilities beginning to be popular at this time. The focus was no longer purely on landscape beauty; now tennis courts, football and hockey fields, and dozens of baseball diamonds were shown. In addition, there were swimming pools with showers to encourage the working-class people—many still without private baths—to be clean.[56] That these facilities, most of which were not related to the water setting, were nonetheless located there can be blamed on topography. The flat expanse of the railroad yard roof and the equally level, newly made land, which happened to be near water, were in fact optimal sites. (Figure 22)

The sensibilities of the local neighborhood also pushed these potentially noisy activities toward the river's edge. The Women's League openly objected to the placement of ballfields and a swimming pool adjacent to the central portion of the park. They suggested their relocation near the abandoned dump at 77th Street, or farther uptown, above West 105th Street. It is unclear if this was an effort to keep outsiders out of the predominately white, upper-middle-class Riverside neighborhood or whether it was genuinely a matter of traffic

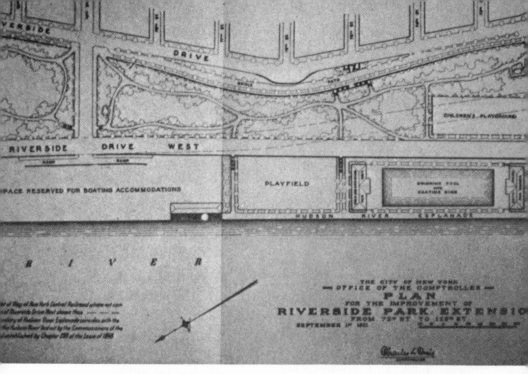

Figure 22. Craig Plan, (1921) 1924. (Copyright © collection of The New-York Historical Society).

and noise. The result was the same: design of parkland became a type of zoning tool that placed new, undesirable social activities as far as possible from private residences.[57]

The automobile, intruding into parks in San Francisco and Chicago during this period, also made its presence known in these new designs for Riverside Park, all of which showed a second thoroughfare. (Map 10) This appropriation of parkland for transportation was deemed entirely justifiable in view of the traffic congestion building up on Riverside Drive. As the smoke and noise of the railroad were on the verge of diffusion, they were replaced by automobile emissions even closer to the homes on the heights. Between 1916 and 1923, car registration in the city tripled. If trends continued, it was estimated that by 1930 there would be one car for every nine residents.[58] Already a cartoon published in a local journal and

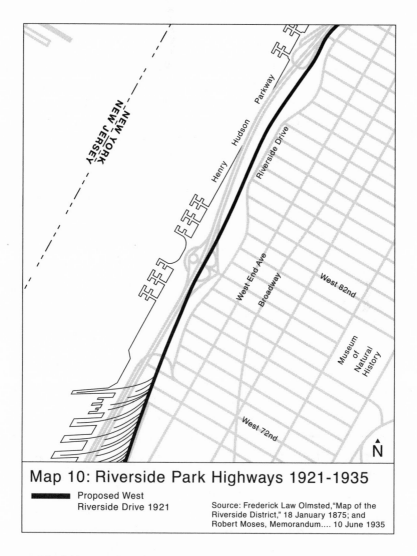

Map 10: Riverside Park Highways 1921-1935

▬▬▬ Proposed West
Riverside Drive 1921

Source: Frederick Law Olmsted,"Map of the
Riverside District," 18 January 1875; and
Robert Moses, Memorandum.... 10 June 1935

entitled "A Midsummers Days Jaunt along Riverside Drive
and Park" depicted the problem and the new technology as
it affected that neighborhood. In one minute from a bench
in the park, one could see the following: a ship, train, horse-
drawn carriage, street cleaning cart, horseback rider, baby
pram, bicycle, double-decker motor bus, and racing car. The

new public works being carried on in the park were seen as an opportunity to create an outlet for the flow of traffic away from the residential neighborhood. Soon there would be opportunities for connections in several directions; the city was beginning to plan for the West Side Highway, and the Port Authority's Holland Tunnel and George Washington Bridge were under way.

The Women's League had an even more elemental perspective on a vehicular route through the park, and once again it was related to local real estate. They and the Riverside neighborhood were still vitally concerned that the railroad tracks be covered. One way to assure this was to press for a highway—strictly for passenger cars—on a roof over the tracks.[59]

After publication of the Craig Plan, discussion of the further development of Riverside Park continued, as in the past, under separate auspices. Engineering matters were addressed by a special committee designated by Mayor James Walker in July 1926, to arrange the removal of surface tracks "from the public streets and avenues on the West Side." Two grade crossings were to be eliminated in Riverside Park in addition to settling the problems of the expanded railroad yards and right-of-way. Membership on this committee was limited to engineers, but they represented the expanding number of agencies now interested in the West Side Improvement. No longer was a group of the Board of Estimate solely in charge; the state, the Port Authority, city transportation, highway, and financing agencies, and the New York Central Railroad Company all sat together.[60]

The Committee of Engineers circulated its *Report* on 13 May 1927, in which it outlined the structural and landscaping obligations of the city and the New York Central Railroad for the yards and tracks in Riverside Park. While placement

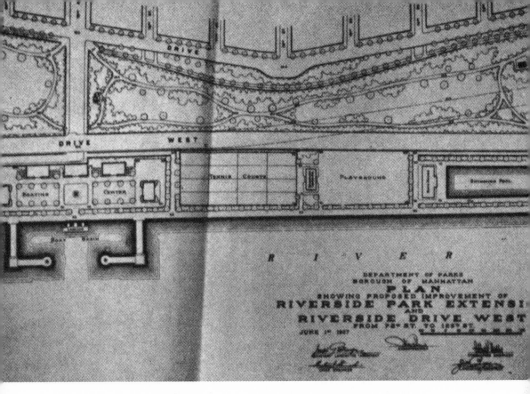

Figure 23. Herrick Plan, 1927. (Copyright © collection of The New-York Historical Society).

of the train facilities was made eminently clear, a landscaping scheme was conspicuously absent. "The Committee has been advised," wrote the authors, "that the Parks Department is now developing plans for the improvement of Riverside Park, but the plans will not be completed in time to permit of their incorporation in this report."[61]

The disconnection of the landscaping or aesthetic aspects of the Riverside Park design from the engineering concerns served once again to make completion of the waterfront park an afterthought of the commercial scheme. The "Herrick Plan" was made public two weeks later, on June 1st. (Figure 23) The Regional Plan Association, a private planning group established two years after the Port Authority also to press for a comprehensive and regional perspective,

143

put the separation in perspective. "The present conditions along the edge of the Manhattan riverside parks," the RPA wrote,

show how extensive are the changes that are being made . . . a "landscaping" plan superimposed on a railroad and street plan, even if prepared by the most skilled of landscape architects, can yield no better result than is secured by the embellishment of an engineering structure, after it is designed, with what is called architectural ornament. Such procedure results in a separation of functions that should be united—and a separation of costs that gives the impression that the architects' contribution to the "design" means an addition to the cost and is a luxury.[62]

At the urging of fine arts groups and civic organizations, the final plans of the decade actually were concerned with aesthetics. After city acceptance of the engineering plans, Mayor James Walker appointed a committee of prominent architects and landscape designers to study the West Side Improvement. Their charge was to refine the artistic elements suggested in the Herrick Plan. Studying these questions were two individuals intensely familiar with Riverside Park, Commissioner Herrick and Albert V. Sielke, an engineer responsible for the Craig Plan, who for six months in 1927 had supervised the placement and grading at Riverside of rocks excavated from subway construction. Among the group of outside consultants was an appointee who reflected the new reality, Herman W. Merkel, Superintendent of the Westchester County Park Commission, which in 1925 completed the region's first parkway.[63]

The discussion in the *Architect's Report* centered mainly on the question of whether the new road in Riverside Park should be on the cover of the railroad tracks or along the

Hudson River. Foreshadowing a dialogue that Robert Moses soon would have with the Borough President of Manhattan, Samuel Levy, there was eloquent rhetoric for both positions. The majority, including Sielke and Herrick, favored a route through the middle of the park in "a beautiful curved line" following the railroad right-of-way. This plan would keep the Olmsted tract intact and allow noisy play facilities to be "as far from the established park as possible." In addition, the city could recapture its waterfront with a riverside promenade "uninterrupted by travelling autos."[64]

The minority pressed for a continuous park covering the railroad, and for a riverfront highway. "The New York Central is about to electrify and cover its tracks," they wrote, "thus giving back to the city its waterfront with all the majesty and serenity which the Hudson River imparts to it. This must not be disturbed by noisy activities [such as a highway through the park]. It is a priceless possession which must imperatively be maintained."[65]

Although the *Minority Report* was defeated at the Board of Estimate, this document provided a platform for constituents favoring a shoreline route. They would shortly come to the support of Robert Moses as he set in motion a financial-design package that after forty-one years would complete Riverside Park six highway lanes short of the Hudson River.

Buttressing the Wall, 1934–1937. The saga of the final years of planning for Riverside Park began to unfold with an agreement between the city and the railroad signed on 2 July 1929. Upon payment of one million dollars, the city now had the right to cover the New York Central tracks from West 72nd Street to Spuyten Duyvil. Work began immediately on the lower half-mile (to 83rd Street), following a design by the

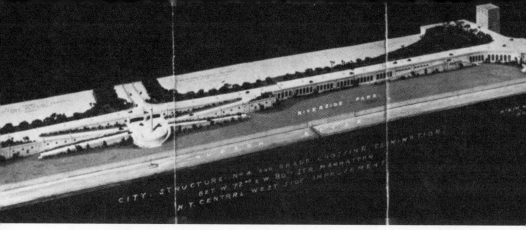

Figure 24. Model for the Railroad Cover, West 72nd to West 80th Street, 1929. McKim, Mead and White, Architects. (Courtesy New York City Municipal Archives).

renowned architectural firm of McKim, Mead and White. (Figure 24) Five years later, Manhattan Borough President Levy initiated construction and financing plans for the roadway above the remaining miles of tracks.[66]

In December 1934 the city's new Parks Commissioner, Robert Moses, questioning the financial and design validity of covering the railroad property, presented yet another plan to the city. Not only was a riverfront vehicular artery included, but once again an ugly scar of open tracks marred the middle of the enlarged park.[67] New hearings were held and tempers flared while Moses and Levy argued over the costs and timing of the two plans. While Moses promised a waterfront highway with a significantly enlarged park for five million dollars, the Levy plan was estimated to cost at least thirty million dollars. The Moses route would relieve the dreadful West Side traffic congestion within a few years, whereas Levy's road would take decades to complete.

Robert Moses even had a rejoinder for the issue of the open railroad tracks. The erection of a series of attractive pedestrian bridges and carefully placed plantings could keep the Olmsted park intact. If the Levy plan were to be followed, threatened

Moses (and this raised the specter of the devastation proposed in 1915), the park would be cut up and unusable for an undetermined period.[68]

On 14 June 1935 Robert Moses set forth his terms for a compromise. In his typical confrontational style, he stated that if his plan was defeated, he would "do nothing more than dump fill north of 83rd Street." However, if the city approved a riverfront highway, he would begin roofing the railroad tracks for park purposes to West 96th Street. A week later, the compromise was adopted (with the tracks to be covered to West 125th Street) at an estimated cost of seventeen million dollars.[69] (Figures 25 and 26)

In the following three years, Robert Moses was able to accomplish what had been under discussion—in over thirty-six unimplemented plans—for four decades. The railroad tracks, reduced to their original two, were covered and the fill (widened fifty feet to the west with debris from railroad excavations below West 59th Street) was landscaped and dotted with recreational features, one of which, the West 79th Street boat basin, even related to the riverside setting. In addition, a parkway that several times had appeared in the Riverside designs became a reality, at the expense of people's access to the Hudson River.[70]

One troubling question lingers: what happened to the power of the Women's League when confronted by the plans of Robert Moses? The group had remained active. In fact, in 1929 it circulated a handbill urging members to write Mayor Walker in favor of the *Majority Report* recommending a highway over the tracks in the park. In typical reform prose they wrote, "to place a highway on the water would deprive mothers and children who cannot leave the city during the summer heat of two miles of cool waterfront."[71] On 27 February 1936 this same group made Robert Moses an Honorary

147

Figure 25. Junction of Riverside Drive and the West Side Elevated Highway at West 72nd Street, 19 April 1939. (Courtesy New York City Municipal Archives).

Member of the Women's League for the Protection of Riverside Park.[72]

One can only guess at what caused the change in attitude. Robert Moses purportedly was a woman charmer. His letters to the league were certainly polite and accommodating. More likely, however, it was his agreement to the one overriding concern for which the league had fought since its inception that tipped the balance. In return for a highway at the edge, he assented to cover the railroad tracks from West 72nd to West 125th Street.

The last luncheon held by the league before it formally

Figure 26. Riverside Park and West Side Highway Landfill, 1937. Sweeny, Harry Jr., ed. *Opening of the West Side Improvement,* 12 October 1937.

disbanded was deliberately scheduled for the fall of 1937. In choosing this date, they indicated their evident pleasure with Moses' work. "Riverside Park," they noted in their minutes, "will be so much improved by that time and we could present our members something really accomplished instead of promises of what was meant to be."[73]

NOTES

1. *Valentine's Manual for 1856*, 594; and William Bridges, *Map of the City and Island of Manhattan with Explanatory Remarks and References*, 25–26.
2. Rodman Gilder, *The Battery* (Boston: Houghton Mifflin, 1936), 120.
3. City Clerk Approved Papers, File "Wharves, Piers and Slips 1835."
4. New York City, Board of Aldermen, *Report of the Special Committee on Parks, Relative to Laying Out a New Park in the Upper Part of the City*, Doc. No. 83, 2 January 1852; and Michael Laurie, "Nature and City Planning in the Nineteenth Century," in *Nature in Cities*, Ivan C. Laurie, ed. (Chichester: Wiley, 1979), 54.
5. Aldermen, *Report on Parks*, 1478; see also: Moehring, *Public Works and the Patterns of Real Estate Growth in Manhattan 1835–1894*, 270–74.
6. Charles H. Haswell, *Reminiscences of an Octogenarian* (New York: Harper and Brothers, 1896), 419; City of New York, Department of Docks, Letters 1870–90; and DD Glass Negative Collection, New York City Municipal Archives.
7. October 1817, 141–42; and Ann L. Buttenwieser, "Awash in New York," *Seaport* (Summer 1984):12–19.
8. State of New York, Ch. 697 L. 1867; and West End Association, *Proceedings*, Doc. No. 1 (New York: n.p., 1871–72), 13. Riverside Park was proposed by William Martin, a Manhattan Parks Commissioner and lawyer for Upper West Side real estate interests in 1865, Stokes, 5:1914. This pamphlet cannot be found.
9. Lillian Wald Papers, File "Parks, Box 28," Manuscript Collection, Columbia University; Morrison, *History of New York Shipyards*, 31, 95, 148, 155, 225; and DD, Letters, File "East 10th Street, 1870."
10. Edward C. O'Brien, "Recreation Piers," *Municipal Affairs* I (September 1897):509–15. By 1910 these piers were located on the ends of the following streets: East 3rd, 24th and 112th streets; Market Slip and Christopher streets; West 50th and 129th streets; and Metropolitan Avenue in Brooklyn. Nathan Straus, who would initiate antinuisance legislation for Riverside Park in the 1920s, founded the free milk program.
11. Document No. 60 of the Board of the Department of Public Parks, "Report of the Landscape Architect Upon the Construction of Riv-

erside Park and Avenue," 1875, quoted in Albert Fein, ed., *Landscape into Cityscape: Frederick Law Olmsted's Plans for a Greater New York City* (Ithaca: Cornell University Press, 1967), 343–48.

12. *Real Estate Record and Builders' Guide*, 22 January 1910, 159.
13. *RRBG*, 5 June 1880, 527. For the history of the physical development of the Upper West Side see: Charles Lockwood, *Bricks and Brownstone, The New York Row House, 1783–1929: An Architectural and Social History* (New York: McGraw-Hill, 1972), 245, 249; Charles Lockwood, *Manhattan Moves Uptown* (Boston: Houghton Mifflin, 1976), 314–18; Moehring, *Public Works*, 279–83, 308–11, 317–19; Hopper Stryker Mott, *The New York of Yesterday: A Descriptive Narrative of Bloomingdale* (New York: Putnam, 1908); and Richard L. Schaffer and Elliot Sclar, "Fashions Seeking a Resting Place: A Century of Change on the Upper West Side," a traveling exhibition of the Division of Urban Planning, Graduate School of Architecture and Planning, Columbia University.
14. *RRBG*, 29 November 1879, 958; and *RRBG*, 5 June 1880, 52.
15. *RRBG*, 18 December 1884, 1251; and *RRBG*, 26 August 1911, 275.
16. DD, Letters, File "West 72nd to 98th Street 1879–90."
17. West Side Association, *Proceedings*, 1871–72, rear cover. According to a contractor who offered to fill behind the bulkhead for free, real estate was not thriving. The contractor withdrew his offer because he was losing money. DD, Letters, File "West 79th Street 1880–85."
18. Land uses at the edge of the Hudson River are found in West End Association and Riverside Park Property Owners Association, "Petition to the Legislature for the Enactment of a Law Providing for the Improvement of the Land and Waterfront Adjacent to Riverside Park in the City of New York," January 1894; West End Association et al., "In the Matter of the Proposed Plans and Agreement between the New York Central Railroad Company and the City of New York Affecting Riverside Park and the North River Waterfront," 3 March 1913, 31–32; Perris and Browne, *City Insurance Map 1873*; Bromley, *Atlas*, 1880; Robinson, *Atlas of the City of N.Y.*, 1890; and "A Prophecy of Twenty-three Years Ago Comes True," Women's League for the Protection of Riverside Park Papers, File "1916," Manuscript Collection, The New-York Historical Society. (Hereafter abbreviated WL Papers.)
19. West End Association, "In the Matter of the Proposed Plans . . . ," 10, 31–35.
20. *Gotham's Great Rotten Row: Peter B. Sweeny's Project for a Splendid Public Pleasure Ground for Lovers of the Horse and the Horse Himself—A Grand Terrace on the West Side . . .* (New York: Municipal Improvement Association, 1890).
21. West End Association, "In the Matter of the Proposed Plan . . . ," 10.

22. Condit, *The Port of N.Y.: A History of the Rail and Terminal System from Grand Central Electrification to the Present*, 140–51.
23. State of New York, Ch. 777 L. 1911. A review of prior legislation may be found in New York, New Jersey Port and Harbor Development Commission, *Joint Report: With Comprehensive Plan and Recommendations* (Albany: J. B. Lyon, 1920), 214–15; and "The West Side Improvement," *The Municipal Engineers Journal* 7 (1921):91–93.
24. Information on landfill is found in New York City, Department of Docks and Ferries, *Joint Report on Proposed Reclamation of Land Between 81st and 129th Streets, North River*, No. 5, 27 December 1910 (New York: M. Brown Printing, 1910); and *RRBG*, 23 September 1911, 415–16. It was also suggested that excavations from the Lexington Avenue subway and street cleaning debris be used; however, there is no record of their deposit.
25. After January 1902, the Dock Board became a single Dock Commissioner. Department of Docks and Ferries, *Joint Report*, Plate 3, 4–6; and *RRBG*, 14 January 1911, 48, 52.
26. *RRBG*, 14 January 1911, 47; *RRBG*, 31 May 1913, 1130; Robinson, *Atlas of the City of N.Y.*, 1890; Bromley, *Atlas*, 1911; and Lockwood, *Manhattan Moves Uptown*, 317. The Potter house is one of the few remaining freestanding Riverside mansions.
27. Annual Record of Assessed Valuation of Real Estate, Riverside Drive, West 121st–22nd Street, 1910. (Hereafter abbreviated Assessments.) One resident, actress Amelia Bingham, had purchased a house at West 103rd Street and Riverside Drive for a low price and had renovated it. She now feared its value would be destroyed. WL Papers, File "Meetings 13 June and 20 June 1916."
28. *RRBG*, 23 September 1911, 415–16; and DD, *Annual Reports 1880–1910*, Rent Rolls.
29. "Scheme for the Embellishment of the Waterfront of the City on its Western Side," *Public Improvements* 11 (1 December 1899):51. For a review of recreation ideology see Galen Cranz, *The Politics of Park Design: A History of Urban Parks in America* (Cambridge: MIT Press, 1982), 177, 283.
30. N.Y. N.J. Port and Harbor Development Commission, *Joint Report*, 215–20; *RRBG*, 31 May 1913, 1130; *RRBG*, 8 May 1915, 777; *RRBG*, 24 July 1915, 131; and *RRBG*, 14 August 1915, 263–64.
31. DD, *Annual Report 1915*, 11.
32. New York City, Parks Department, *Annual Report 1915*, 22. Three years earlier Stover had written to Mayor Gaynor: "I believe the time has come for the Board of Estimate and Apportionment to allow Parks money for surveys . . . and the preparation of plans for the improve-

ment of the entire front between 72–129th Street," 14 December 1912, Mayors Papers, File "Gaynor, Parks 1912," New York City Municipal Archives.

33. *RRBG*, 29 April 1916, 654.
34. West End Association, "Report on proposed deal between N.Y. Central R.R. and City of N.Y. relating to Riverside Park and North River," 4 May 1916, 3. The open dump at West 79th Street had been moved to an enclosed building at West 77th Street.
35. Material for this section was found in WL Papers; and WL Scrapbooks, The New-York Historical Society; *RRBG*, 13 May 1916, 719; and *RRBG*, 2 September 1916, 327.
36. "Whom They Endorse," *Report of the City Club of New York*, 24 June 1916; Citizens Union, "Memorandum Concerning Proposed Relocation of New York Central Railroad Tracks Upon the West Side of Manhattan," April 1916; and Letter from Municipal Arts Society to WL, 12 June 1916, WL Papers, File "1916–17."
37. WL, *By Laws*, WL Papers, File "1916"; and WL Scrapbook, 17 October 1916. (author's italics)
38. William Bonner, *New York: The World's Metropolis 1623-4—1923-4*; and WL Papers, File "1916." Nanette Bryan became president of the league in 1918. Mrs. Kerr's father, Andrew R. Culver, tried unsuccessfully to develop Coney Island "along the lines of Brighton." He eventually sold out to the Long Island Railroad. WL Papers, File "1916."
39. Bonner; *New York:* and Trows, *City Directory*, 1920.
40. Letter to Hon. Frank L. Dowling, President, Board of Estimate and Apportionment, 4 August 1916, unsigned, WL Papers, File "1916 D."
41. "Whose Child Next?" WL, Scrapbook, undated; and WL Papers, File "1916, 1917."
42. WL Papers, File "1917" including handwritten notes on the history of the league's first year, 6.
43. "Women to Keep Up Fight," 31 January 1917, Mayors Papers, File "Mitchell Scrapbook," New York City Municipal Archives. Details of the Women's Movement during this period are found in William H. Chafe, *Women and Equality* (New York: Oxford University Press, 1979), reprint ed., 12–13, 16–17, 28–29; and Ann D. Gordon, Mari Jo Buhle, and Nancy Dye, *Women in American Society* (Somerville, Mass.: New England Free Press, 1972), 44. The role of women in planning is found in Eugenie Ladner Birch, "From civic worker to city planner: women and planning, 1890–1980," in *The American Planner*, Donald Krueckenberg, ed. (Cambridge: MIT Press, 1983), 393–426.
44. Jens Jensen, "Report to the Women's League for the Protection of

Riverside Park on the Proposed Plan for Changes in the New York Central Railroad Along the Hudson," Chicago, 26 November 1916, 10, WL Papers, File "1916."

45. WL Papers, File "1916, 1917."
46. Correspondence between Frances Peters and William Prendergast, January 1917, WL Papers, File "1917 P"; Letter to Dowling, 4 August 1916, WL Papers, File "1916 D"; Interview sent in answer to Dock Commissioner Smith's abusive letter of March 30 to *New York Post,* WL Papers, File "1916"; and Mitchell Scrapbook, December 1917.
47. WL Papers, File "1917."
48. Shortly before the league held its first mass meeting (May 1916) Marks suddenly parted company with the other members of the Ports and Terminals Committee and announced that he opposed the railroad plan. His reasons for this decision are unclear. However, news of his defection was sent to the Women's League by reformer-lawyer-historian Frank Moss, and Marks was promptly invited to speak at the mass meeting. BP *Annual Report,* 1916, 44–45; and WL Papers, File "1916."
49. Mrs. Julius Henry Cohen, "The West Side Problem Nearing a Solution," *Women and the City's Work* 3 (23 October 1917).
50. "A. A. Taylor Street Cleaning Commissioner Opposes Bill to Remove Dumps," WL Scrapbook, 14 February 1922; WL Papers, File "1923 Sanitation Committee"; and *RRBG,* 8 May 1915, 773.
51. DD, Letters, File "West 72nd–129th Street, 1920–25." Known as "Hylan's Folly," the Staten Island complex was little used. DD, *Annual Reports* 1920–30; and N.Y. N.J. Port and Harbor Development Commission, *Joint Report,* 175. The value of exports and imports of the port of N.Y. as a percentage of the national total declined as follows:
    1900—49.6 percent
    1913—46.2 percent
    1923—41 percent
    RRBG, 10 April 1915, 598; and *Fortune* 20 (July 1939):158.
52. WL Minutes, October 1930; and WL Papers, File "1923–30."
53. John R. Slattery, "Modern Methods of Subway Construction," *The Municipal Engineers Journal* 12 (1926):173; and Memo to the Board of Estimate and Apportionment from Board of Transportation, 5 February 1927, Mayors Papers, File "James J. Walker, West Side Improvement," New York City Municipal Archives.
54. Open letter from Mrs. Charles Austin Bryan, undated, WL Papers, File "1922"; and Charles L. Craig, *Riverside Park Improvement* (New York: n.p., 30 January 1924). It is unclear whether this is a revival of a Parks Department Plan that was promised in 1917. The map was

drawn by Parks' staff in 1921. Resume of Albert Sielke, Stanley Isaacs
Papers, File Box 2 "1937 Sielke," Manuscript Collection, New York
Public Library.

55. Assessments, 1910, 1923, Block Nos.: 1185, 1240, 1250, 1276.
56. Cranz, *Politics of Park Design,* 70, 87–88.
57. Letter to Mayor Walker, 10 June 1926, WL Papers, File "1926 W";
    and Robert A. Caro, *The Power Broker: Robert Moses and the Fall of
    New York* (New York: Knopf, 1974), 512–14. According to Caro,
    Moses purposely filled pools in white neighborhoods with cold water
    to keep blacks from swimming there.
58. Ernest P. Goodrich and Harold M. Lewis, "The Highway Traffic
    Problem in New York and its Environs," Regional Plan Association,
    Bulletin, May 1924; and Armand R. Tibbitts, landscape architect,
    "Statement to the Women's League on Proposed Riverside Improve-
    ments," 17 April 1926. "Regarding the [driveway over the tracks] as
    an artery of traffic," he wrote, "I have nothing to say except that
    without a doubt it is badly needed." The cartoon was drawn by Jef-
    ferson Machamer, WL Scrapbook, 1924.
59. WL Papers, File "1924–29."
60. The following were members of this committee: R. E. Dougherty,
    New York Central R.R. Co.; William C. Lancaster, Transit Com-
    mission; C. M. Pinckney, Borough President's Office, Manhattan; J.
    R. Slattery, Board of Transportation; Arthur S. Tuttle, Board of Es-
    timate and Apportionment; and Billings Wilson, Port Authority of
    New York and New Jersey.
61. *Report of the West Side Improvement Engineering Committee,* 13 May 1927,
    11, Mayors Papers, File "Walker, West Side Improvement," New
    York City Municipal Archives.
62. "The West Side Waterfront on Upper Manhattan," March 1928, 5.
63. Mayors Papers, File "Walker, West Side Improvement."
64. *Report of West Side Improvement Architects Committee,* 29 April 1929, 3,
    Mayors Papers, File "Walker, West Side Improvement."
65. *Report of West Side Improvement Architects,* 8.
66. Material for this section was derived from Caro.
67. *NYT,* 16 December 1934, 11:2.
68. *NYT,* 26 April 1935, 23; and Board of Estimate and Apportionment,
    "Miscellaneous Hearings Borough of Manhattan No. 6," 24 May 1935,
    7–8, WL Papers, File "1935."
69. *NYT,* 14 June 1935, 25, 55. Caro, 556–57, reports that Moses estimated
    the cost at twenty-four million dollars, but that expenditures on the
    entire West Side Highway were actually $180–218 million. If one de-
    ducts the elevated section downtown (seventy million dollars), then

the drive from West 72nd Street north cost at least $110 million, or four times the original estimate.

70. Letter from Lillian Francis Fitch to Jean Kerr, WL Papers, File "1927 F."
71. Letter from Ann E. F. Ryan to Mayor James Walker, 18 March 1929, Mayors Papers, File "Walker, West Side Improvement."
72. WL Papers, File "1936 M."
73. WL Minute Book, February 1937, WL Papers, File "1936–38."

CHAPTER 6

Thickening the Wall

A new structure began to line the waterfronts of America after 1930—the arterial highway. Although they were designed originally as luxuriant boulevards and the focal point of a day's outing, the vision of these roadways was quickly transformed into a resource to relieve cities of congestion and blight. From Lake Shore Drive in Chicago to the Harbor Freeway in Los Angeles to the elevated arterials of the 1950s in Boston and Seattle, the effect everywhere was not only to accommodate cars, but also to restrict human access to the water.

New York was not unusual in considering waterfront roadways. Daniel Burnham's 1905 plan for San Francisco called for an outer boulevard skirting the bay, with plantings and viewpoints where pedestrians could watch the shipping. His *Plan of Chicago*, published three years later, envisioned a "supremely beautiful parkway" along the shores of Lake Michigan, with fields and playgrounds to attract not only the inner-city working classes but "people of means" who would have vacation homes there. Landscape architect John Nolan recommended a similar "Harbor Drive" for San Diego that would function on weekdays as a "bypass route around congested areas of the downtown," and on evenings and weekends as a pleasure drive.¹ In New York Nelson Lewis, Chief Engineer of the Board of Estimate and Apportionment, described how such a roadway would work:

There is offered a remarkable opportunity for an elevated parkway similar to the one on the Algiers water front. The trucking traffic could go on beneath, as it does now, and on the upper level there would be ample space for motors and busses to speed to the Battery unhampered by cross traffic. Occasionally this Water Front Parkway, with frequent ramps from upper to lower levels, could be widened into small parks and playgrounds, with salt water swimming pools for the children, thus providing generous recreation facilities for the densest population in the world which parallels the river all the way but is now practically shut off from it. . . . Eventually this Water Front Parkway could be extended all the way around Manhattan Island. Burnham's dream of getting folks to water all over again!²

Roads Along the Wall, 1890–1915. The first segment of New York's arterial system was actually begun in 1893, when the Department of Parks was authorized to lay out the Harlem River Speedway. The drive was modeled after Sweeny's *Rotten Row*, the harness-racing circuit proposed a few years earlier for the Hudson River edge of Riverside Park. The two-and-one-half miles of Harlem River racetrack ran, uninterrupted by side streets, from East 155th Street to Dyckman Street (198th Street). (Map 2)

As described by the Parks Department after its opening in

1898, the speedway assisted the horse-breeding industry. Used chiefly for trotting horse trials by New York's elite, it was "intended to be a resort of those drivers who desire a specially prepared roadway for the best class of driving horse-drawn vehicles. . . . " No carriage or business carts were permitted, and it was the only public drive "with no speed restrictions."[3]

Although the "attractions of a riverfront" were promoted for the roadway and for the accompanying pedestrian walks used by racing spectators, there were other reasons for its location. Because it was waterfront land, most of it belonged to the city; therefore, no purchase was required. Here too was an uninterrupted stretch of land that, due to the steepness of its slopes, would be difficult to develop for commerce or real estate. The remoteness of the area was also desirable. Unlike the crowded riding paths of Central Park, here was room for a sizable and affluent constituency of riders to enjoy their sport undisturbed.

The use of the speedway for recreation was short lived. Less than a decade after its opening, north-south traffic congestion and the desire of realtors to open upper Manhattan prompted calls for its extension westward, across Dyckman Street to the Hudson, and for its use by commercial vehicles. In 1909 the Washington Heights Taxpayers' Association began lobbying for redesignation of the speedway as a through route, not for horses but for cars.[4]

While the waterfront speedway was adjusted to handle increased traffic uptown, the throngs of people and vehicles downtown were reaching a saturation point. Forecasters of the city's population estimated an increase of a million persons by 1920. In addition, new bridges, tunnels, and steam railroads would bring many of these individuals into Manhattan, where the competition to construct the tallest building was just beginning. The street system, designed for entirely different conditions, would soon be totally inadequate. Already pedestrians feared for their lives and business suffered because neither goods nor buyers could reach shops.[5]

Nineteenth-century complaints had been of lines of horse-drawn carriages, carts, and railways blocking busy thorough-fares; the automobile, wending its way through newly created canyons of skyscrapers and between stanchions of the elevated, was the twentieth-century threat to the quality of life and business in New York. The *Real Estate Record and Guide* sounded the alarm in 1908, claiming that the automobile would soon overload New York's streets. "The constant process of improvement and cheapening that is taking place in motor-cars," editors asserted a year later,

will result in an increase in vehicular traffic greater in proportion than the increase in population. If it is difficult for the street system of the city to accommodate this traffic now, what will it be fifteen or twenty years from now, when that traffic will have at least doubled in amount . . . the crowds of vehicles and people on [Manhattan] streets . . . will produce a condition of congestion which will gradually become intolerable.[6]

At fault, according to urban critics, was the city's first plan, the Randall Commission's Map of 1811. Its first mistake, according to the Municipal Art Society, "lay in projecting comparatively few north and south avenues . . . and inordinately numerous east and west streets which are not so necessary."[7] In a 1915 article entitled "City Planning," Nelson Lewis discussed the failures of the Randall Commission to allow motion not only within, but in and out of the city. "With the expansion which has taken place in all directions," wrote Lewis, "the few north and south avenues are so overtaxed that, notwithstanding the fact that transit lines have been built over . . . and under . . . the need of additional thoroughfares in this direction is quite apparent, and the almost entire absence of diagonal or radial highways prevents direct access from one side to the other at points above or below."[8]

Modernization of the interior street system and new roadways along the waterfront were among the solutions proposed. Wider and longer north-south avenues and new diagonal streets were constantly discussed. Double-decking was another expedient. In 1907, Charles R. Lamb, President of the Municipal Art Society, recommended that the pedestrian sidewalks of the east-west streets leading to ferry terminals be elevated, freeing the ground level for commerce. The New York City Improvement Commission, in its endorsement of the ornamental design of the Chelsea Piers, pressed for an elevated street between the piers and where the line of buildings began, "built along the waterfront to accommodate the North and South travel. . . . Approaches might extend from this elevated street to the recreation piers. . . . Stairs for the descent of foot passengers and approaches by inclined planes for horses and vehicles can be made at convenient places."[9]

The West Side Improvement Downtown, 1910–1930. Between the 1906 passage of the Saxe Law, requiring the elimination of steam railroads from grade on the streets of New York, and the 1918 postponement (with the help of the Women's League) of any New York Central development from Spuyten Duyvil to St. John's Park, the Lower West Side of Manhattan became the prime arena for traffic improvement plans. Here, without the tracks, would be newfound space that could be improved for commerce. Ernest Poole, in a 1915 novel entitled *The Harbor,* described the problem: "On the Manhattan side of the North River, from Twenty-third Street down for a mile there stretches a deafening region of cobblestones and asphalt over which trucks by thousands go clattering each day. There are long lines of freight cars here and snorting locomotives... along the water side is a solid line of dock-sheds. Their front is one unbroken wall of sheet iron and concrete."[10]

The majority of schemes to reorder the Lower West Side, reminiscent of the 1870 proposals to the Dock Department, were concerned with belt-line railroads, and with terminals and warehouses for moving and storing freight. A rather modest proposal was put forth in 1908 by W. J. Wilgus, who recommended a two-track freight subway that encircled the island under the streets bordering the rivers. The subway would have access to the basements of adjoining buildings. A later suggestion expanded the Wilgus idea to sixteen peripheral tracks.[11]

Public officials, also vitally concerned with solving the problem of the "West Side Improvement," had broader views. In a series of proposals issued between 1910 and 1912, Dock Commissioner Calvin Tompkins sought a comprehensive freight system including an elevated railway, six terminals, and multileveled storage and transfer yards. All were to be located below 60th Street between the Hudson River and Tenth Avenue. The New York and New Jersey Port and Harbor Development Commission, forerunner of the Port Authority and the first official regional planning body in New York, devoted two chapters in its *Joint Report* of 1920 to a debate over the merits of an elevated versus a submerged freight route encircling Manhattan.[12]

The most imaginative proposal for a waterfront transportation system was offered by T. Kennard Thompson, consulting engineer on over thirty skyscrapers, including the Singer Building. He was also noted for an earlier Jules Verne-like proposal to extend Manhattan by fill to Governor's Island. In a scheme that recalled the J. Burrows Hyde design of 1870, Thompson recommended enclosing Manhattan with a six-track, belt-line railroad to run at grade along the outer edge of the island. On top would be structures that accommodated motor vehicles. A half-block-wide elevated local street would connect via ramps to the second story of adjacent buildings. Above this, a narrower speedway would enable cars and trucks to cruise uninterruptedly for a minimum of a mile. Although Thompson's plan was praised as employing "in the most effective and utilitarian manner the great, broad, open highways that face Manhattan's waterfront," it remained, as the other plans, an idle idea.[13]

Postwar prosperity allowed more people to purchase cars in the mid-1920s, adding to the congestion on the main avenues of the island; and the still unsolved railroad problem was once again under negotiation. Discussion of an elevated roadway along the lower Hudson River shoreline thus began anew. A river route was deemed appropriate because together, the through and service streets provided enough width to hold the necessary stanchions without invading the property reserved for port activities. In addition, once cars were diverted to an upper level, the heavy trucks associated with maritime uses would be able to move more expeditiously on the existing, repaved, outer street.[14]

Land along the water was also significantly cheaper to acquire. Public ownership would provide some savings. The decline in private property values along the West Side waterfront that had been evidenced in the negotiations for the Chelsea-Gansevoort improvement, and that persisted in the 1915 discussions over the New York Central track removal,

163

provided additional reductions. Contrast the costs of acquiring land for widening inner-city avenues with acquisition at the western edge and the idea of a waterfront highway becomes even more compelling. Extension of Sixth Avenue through private property from West 3rd Street to Canal Street in the 1920s cost $10 million per mile. In midtown the figures would have been at least quadrupled. Alternatively, at the Hudson River between West 60th and Canal streets, land averaged $4 million per mile, or less than half of the Sixth Avenue elongation.[15]

The issue was finally put to rest by the New York State Legislature in 1926, when it authorized construction of a roadway on a viaduct from Canal Street to West 72nd Street. (Figure 27) Named the "Miller Highway" for the Manhattan Borough President who conceived and fought for it, the elevated West Side Highway was immediately praised in the press, which asserted that, "the Driveway will furnish a continuous view of the river and the shipping."[16] There was no mention of the pier sheds and highway railings that would interrupt that view, nor of the visual barriers the highway itself would create as it crossed dozens of east-west streets. (Figures 28, 29)

The highway was not without its critics. In a memo to an assistant to Mayor Walker, Daniel L. Ryan of the Board of Transportation asked: "With the tracks of the New York Central Railroad removed from the surface of both 10th and 11th avenues . . . why is there a need for a high-level, high-speed highway? . . . there is no reason why 12th Avenue should not be opened the entire length of Manhattan Island as far as 58th Street . . . to provide a through highway, such as the Hudson River front is certainly entitled to, all the way to New York Central yards."[17] There were also those who questioned whether the new highway, in relieving the inner

Figure 27. Design Made to Show Finishes for the Elevated West Side (Miller) Highway. Sloan and Robertson, Architects, c. 1926. (Courtesy New York City Municipal Archives).

streets of congestion, would not succumb to immediate over-crowding. In a clairvoyant statement to the Board of Estimate and Apportionment shortly after the state authorized con-struction, Chief Engineer Arthur Tuttle wrote: "While a roadway of this capacity may appear adequate for the re-quirements, I am firmly convinced that within a very short period after its completion the proposed structure would be over-crowded and its efficiency seriously crippled by reason of the necessity of sacrificing speed to provide for an excessive volume of traffic."[18]

Figure 28. Waterfront View from West 22nd to West 24th Street, from the West Side Highway in Construction, 1932. (Courtesy New York City Municipal Archives).

When the first sections of the West Side Highway were opened in 1931, plans were already under way for its extension along the East Side. Although the initial route was prompted largely by the chronic congestion that for over half a century had plagued West Street and Twelfth Avenue, the East Side experience was different. Here, where shipping activity was drastically reduced and living conditions were a residential Hades, the East River Drive became a slogan for slum removal.[19]

166

Parkway and Luxury Housing, 1929. In the blocks of the Lower East Side south of the contemporary East River Park, west of Pier 44 and north of Gouverneur Hospital, known as Corlears Hook, were some of the most decayed real estate to be found in New York City in the third decade of the twentieth century. (Map 11) Dilapidated commercial buildings and boarded-up warehouses dotted the land. Most of the property consisted of dank, decrepit, and rat-infested old-law tenements. Over three-quarters of these buildings lacked heat and contained only hallway toilets. The East River waterfront was lined with dumps, coal, and lumber yards, and the vestiges of a once-flourishing shipbuilding industry—abandoned machine factories and rotting piers.[20]

Planning for the physical rejuvenation of the Lower East Side waterfront was initiated not by the Dock Department, the agency in charge of the city's outer streets in prior years, but by the Borough President of Manhattan. This office had been given exclusive jurisdiction over these streets in an effort to alleviate the delays in development that had allegedly been caused by joint control with the Dock Department. Julius Miller therefore became the city's first parkway entrepreneur. In January 1929, with his West Side Highway in progress, politician and lawyer Miller presented a plan for an East River Drive to the Board of Estimate.[21]

Once again a waterfront location was chosen for the city's newest north-south thoroughfare, and the decline of shipping along the Manhattan side of the East River provided an incentive for this choice. The opening of bridges and tunnels beginning in 1927 and an increased use of trucks for transportation had reduced the demand for railroad lighters and ferries. Canal barges, a few steamship firms, fishing and oyster boats, and some railroad companies transported food and freight to and from aged piers below Grand Street; and the

Figure 29. View South from the Elevated West Side Highway Showing the Chelsea Piers from West 19th Street, 1930. (Courtesy New York City Municipal Archives).

only remaining passenger ferries were at Whitehall, East 29th, and East 125th streets. A municipal ferry also transported the dead from East 78th Street to a pauper's burial ground on Hart Island. Above Grand Street some of the remaining East River landings were used by industrial holdouts—power plants, breweries, laundries, and slaughterhouses—many of which sat adjacent to the growing luxury residential section between Kips Bay and Sutton Place (East 27th to 59th streets). One pier around Old Slip housed a new waterfront activity—

168

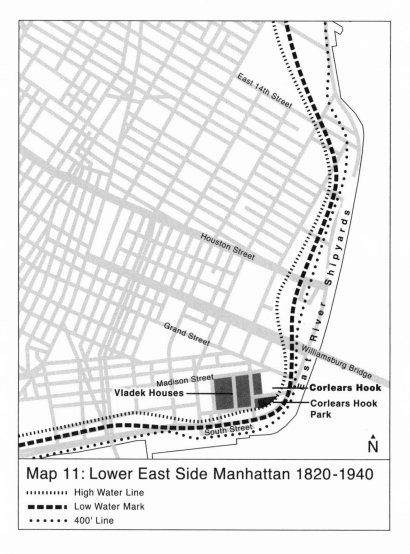

East 14th Street

Houston Street

East River Shipyards

Grand Street

Williamsburg Bridge

Madison Street

Vladek Houses ——

Corlears Hook

——Corlears Hook
Park

South Street

▲
N

Map 11: Lower East Side Manhattan 1820-1940

▪▪▪▪▪▪▪▪▪▪▪ High Water Line
▪▪▪▪▪▪ Low Water Mark
• • • • • • 400' Line

a seaplane base—and familiar, old barges carrying building materials, fuel, and ice begged for space along the rest of the shore.[22]

The river route in addition allowed an ingenious method of financing that might reduce land acquisition costs. Where the proposed roadbed crossed private land, Borough President Miller intended to revive the eighteenth-century water grant clause that had required owners adjacent to the rivers to build a portion of the outer street. In return for relief of this obligation, which historically owners rarely had honored, the city hoped to reclaim the land at no cost and thus save seven million dollars.[23]

Although the newest link in the city's arterial system was planned for the entire length of the East River in order to connect to the West Side Highway on the southwest and the Harlem River Speedway on the north, construction was slated to begin on the Lower East Side. Here, a wide, shoreline roadway would serve a section that, because of a bulge in the island, was nearly a mile from any direct line of transportation. The closest north-south bus routes were on avenues B and C, while the nearest elevated was on First Avenue, three long blocks to the west of the bus line and a mile from most users.

Cheaper property and less complicated ownership patterns along the East River also induced development to begin there. A parcel near Henry Street, four blocks from the shore, was valued at $7.50 a square foot in 1931. The abandoned Crane factory opposite Corlears Hook Park, which bordered the water, was two dollars lower. On the interior blocks, furthermore, each lot was often owned by a single family or estate; this fragmentation made the accumulation of large, contiguous plots of land extremely difficult. At the island's edge the Corlears Hook Park waterfront and piers from Grand to East Houston Street were already city property. The nine riverside blocks between East 4th and 13th streets were held by only five owners.[24]

There was in addition an aesthetic justification for a waterfront location. The Lower East Side, "The Mother of Urban Slums in the United States," was ripe for renewal.[25] The success of Frederick Law Olmsted, Burnham, and Lewis in upgrading waterfronts in New York and other American cities led many city planners to believe they had found a solution to a major urban problem—the obstacle that commercial and nuisance activities presented to the creation of higher-income housing. In New York, it was suggested that an attractively landscaped East River driveway would lead to slum removal inland.

The road thus became a popular public relations device for higher-income housing. Borough President Miller in his *Annual Report* for 1929 set the tone. "The construction of the . . . Drive," he wrote, "will open up the entire abutting territory to high class development, both commercial and residential, and will beautify over three miles of East River shore front." The roadway's transportation purpose was left to the end of the sentence, "as well as provide a much needed additional north and south traffic artery."[26]

Proponents and Their Plans, 1930–1933. Initially, the most vocal proponents of the drive were local property owners who would gain from development of the area for a more lucrative use. A committee of prominent citizens was organized by the Greater New York Taxpayers' Association. It included representatives of the Merchants' Association, savings banks, Beth Israel Hospital, and the renowned Rabbi David de Sola Pool. Testifying at a Board of Estimate hearing in 1930, the committee's representatives affirmed that the roadway "would open up to intensive apartment house development an area rapidly declining."[27] *The Real Estate Record and Guide* hinted at the quality of housing: "[The Drive's] greatest promise lies in the effect it might have on residential development along the East River. What already has happened in the Sutton Place section and in the vicinity of Beekman Place might be duplicated on a much greater scale at other points along the proposed Drive."[28]

The goals of the Lower East Side Chamber of Commerce were at once broader and conflicting. The group favored the roadway for its financial promise. Within brisk walking distance was Wall Street and its upper-income jobs, which with the road would bring in a higher class of residents. Yet concurrently, the Chamber extolled the drive's social role as open space that would benefit the existing low-income community. The group envisioned not a highway for commercial vehicles, but a "beautiful parkway" along a "picturesque East River waterfront" that would have a beneficial influence on the health and behavior of current, low-income, local residents.[29]

The incongruity of these goals reflected the wishes of a diverse membership. The Chamber comprised not only local savings banks and businesses that would benefit from upgrading their properties, but local settlement houses and religious institutions that had a social mission. Despite their differences, these groups shared a common belief or naïveté that somehow the poor would not be displaced by luxury activities that would upgrade the area.

Objections to the first section of the drive came from professional planners, whose resistance related more to the one-sidedness of the Miller proposal than to its substance. Among the initial requests of New York's first City Planning Commissioner, John F. Sullivan, appointed by Mayor James Walker in 1930, was to delay the approval of the East River Drive. He wished, in consultation with the Borough President and the Dock Department, to draw up a comprehensive development plan for the entire Lower East Side. Together, Sullivan hoped, city officials would consider multiple urban problems: traffic relief, slum rehabilitation, recreation, and heavy freight-loading.[30]

In 1929 the Regional Plan Association (RPA) began publication of a multivolumed survey of the physical and social facilities in the New York region, together with a plan for its future development.[31] Detailed studies of various sections of the city were undertaken too, and in 1930 the RPA issued two reports on the East Side waterfront. The first, a development scheme for Manhattan's shore north of East 23rd Street, shows how far planning for the city's waterfront had evolved. In a proposal never intended for implementation, the architects of the RPA depicted a monumental, multi-leveled row of apartments, offices, roadways, garages, and public buildings lining the water.[32]

This report also reflected the tension between commercial needs and civic and residential desires that had existed at Riverside a decade earlier. Nuisance activities such as dumps and slaughterhouses should go, it insisted, while certain water-dependent activities could be accommodated. Electric power stations, which had replaced the gashouses, could be masked with "special architectural treatment," and shipping could remain because residents would not object, and "there is nothing more interesting than to drive along a fine marginal way and witness the activities connected with navigation."[33]

The RPA's second publication, the *Lower East Side Report,* was an attempt to reconcile the Miller highway design with the confusing notions of local real estate groups and social agencies. As a further incentive to high-class residential investment, the planners realigned the highway to include an enlarged park specifically for the wealthier newcomers. An inland drive would follow Mangin Street (two blocks to the west of the river) from Montgomery to East 3rd streets, and then run north at the water's edge. Along the outside of the roadway, below East 3rd Street, was an elongated Corlears Hook Park "sufficiently large to attract *high class* residence" (author's italics). A yacht basin would frame both ends of the open space.[34]

The RPA's conception was contradictory. It recommended high-class housing along a drive that was to be part of a "highway for uninterrupted traffic circling Manhattan," hardly a use compatible with luxury residences. In addition, it outlined at least four east-west, Park Avenue-type boulevards (three blocks apart) that "make the waterfront accessible to the inland area and thus raise property values."[35] It is unclear whether the poor from the inner city would thus have easier ingress to the park or whether luxury housing would obviate this need by extending inland on the new avenues.

In 1931 several social reformers also expressed support for a complex of highway, park, and upper-income housing for the Lower East Side. In an article written for *Survey,* the Journal of the Charity Organization Society, Loula Lasker (sister of philanthropist Albert Lasker) commented on the Miller and RPA plans: "All agree that the East Side must look for its regeneration to an influx of a population of a much higher scale." With improvements, she felt, the poor would be able to filter into housing left vacant in uptown Manhattan by the exodus of the rich to the Lower East Side. Instead of having the rich move farther out onto the undeveloped periphery of the city, as was the pattern in the past, Miss Lasker was recommending a reverse process similar to gentrification.

The waterfront, Miss Lasker wrote, was a "really priceless asset" and it could become "the most potent factor" in the Lower East Side's revival. She also voiced the first note of caution in all of these statements, a concern that the proposed motor highway would simply cut off those living on the interior of the Lower East Side from the city's shore. Yet she stopped short of carrying her observation to its logical conclusion. Citing the example of Chicago's Lake Shore Drive, she wrote, "... even an oblique view of the river through a window has a marked effect on increasing rents. Is New York less resourceful?"[36]

By this time the Lower East Side Chamber of Commerce was actively pressing for white-collar housing. It hired consultants, among them architect Arthur Holden, to conduct land use surveys and to devise a master plan as a base for future change. This framework included not only major thoroughfares but suggestions for upper-income residences, business and recreational development, and rezoning of the waterfront from unrestricted to residential.[37]

Plans for the drive were meanwhile refined reinforcing it as the setting for quality housing, but there was very little construction activity. In October 1931 Borough President Samuel Levy, Miller's successor, submitted drawings to the Board of Estimate for the first section of the East River Drive. The promotional material elicited visions of an elegant avenue rather than an arterial highway. A four-lane boulevard was bordered on the west by a sidewalk, trees, and high-rise apartments. On the east, additional plantings and three acres of parkland screened a service street and the cleaner commercial activities that would be allowed to remain on refurbished piers. Local east-west streets crossed both the commercial and through roadways.[38] (Figures 30, 31) (Map 12)

Two years later in revised plans for the drive, Borough President Levy added elements that had been effective in attracting luxury housing to Riverside Drive. A "wide parked street" was to be extended south to the Brooklyn Bridge, and decaying piers would be replaced by a thirty-five-acre

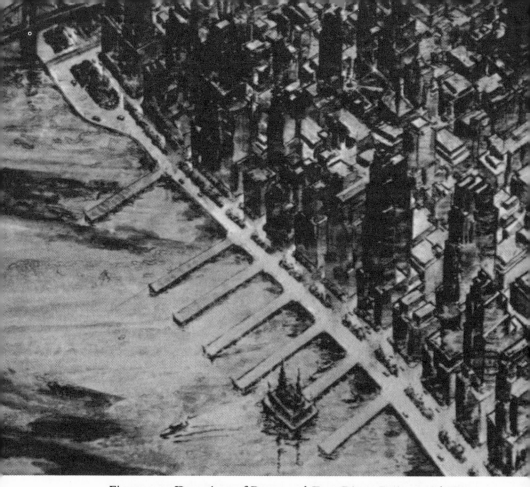

Figure 30. Drawing of Proposed East River Drive with Recreation Pier Structure, c. 1931. (Courtesy New York City
Municipal Archives).

East River Park. Below East 30th Street, although it was to
be six lanes, Levy's highway was still a boulevard with connections to every local street and trees lining the center
divider.[39]

Housing for Whom? 1933–1934. The Chamber of Commerce's
call for a Lower East Side master plan sparked the imagination

176

Figure 31. Drawing of Proposed East River Drive and Housing, c. 1934. (Courtesy New York City Municipal Archives).

of another architect, Andrew Thomas. In 1933 he reintroduced a concept that had been tried here before: the provision of better low-income housing. Thomas came with good credentials. Earlier, he had designed both new-law tenements and Harlem's Dunbar Apartments—the latter for John D. Rockefeller. He now unveiled for the Chamber his vision of a massive slum clearance project in the 135 blocks between the Manhattan and Williamsburgh bridges. In a drawing reminiscent of the earlier RPA grand scheme, he showed a mul-

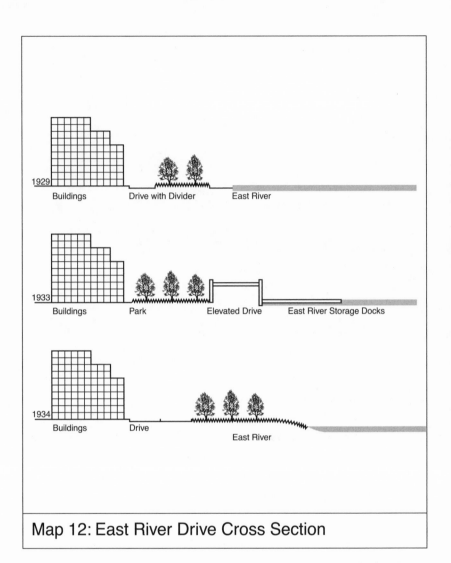

Map 12: East River Drive Cross Section

tilayered shore, with highway and park elevated two stories above the existing outer street. Underneath, in one more effort to revive the commercial waterfront, yet carefully hidden from view, were piers and a service street.[40] Attached to the west were offices and garden and high-rise apartments. Over half of these residences would be rented to middle-income tenants paying monthly from nine to twelve dollars a room. The remaining units would be marketed as low-income housing at $7.50 a room per month. While these rents may have appeared low to Thomas, these units would not have housed many existing Lower East Side residents. In the Corlears Hook section, for example, a quarter of the families paid under four dollars monthly for a room.[41]

The Thomas plan reflected a lingering pressure from coal businesses in the area to protect their yards and landings; it also represented a growing constituency for low- and middle-income rental units. While Andrew Thomas was peddling his proposal, realtor Robert W. Aldrich Roger was actually seeking private funding for a smaller, low- and moderate-income project, "Rutgers Town." It would be built on eighteen blocks of cleared slums below Corlears Hook Park between Jackson and Gouverneur streets, Broadway and the East River. As in Thomas' scheme, monthly rents were scheduled at an unrealistic six to twelve dollars a room.[42]

By 1933, the suggestion that upper-income housing would be viable on the Lower East Side appeared less likely. The Thomas and Rutgers Town plans were one reflection of a declining demand for high-income rental units. Expanding luxury development on the Upper East Side contributed to this shift. Tudor City at East 42nd Street, Sutton Place at East 53rd Street, and River House and the Campanille at East 52nd Street were all under way. A report from the Lower East Side Chamber of Commerce pointed out the limited market: "it would require that all the families in Manhattan and half the families in Queens [paying] fifty to one hundred dollars a month rent in 1934, move to the Lower East Side in a body. Manifestly, this is impossible."[43]

These plans were also evidence of how firmly entrenched low-cost housing was perceived to be in this area. As early as 1833, James Allaire, owner of a nearby marine engine-works factory, established New York's first four-story tenement house at Water Street and Corlears Hook. In the decades that followed many more tenements were built, and owners sought greater profits by cramming more people into smaller spaces. An increasing number of immigrants raised residential densities from 54.5 persons per acre in 1820 to 450.2 in 1870. Landlords of four-story tenements multiplied their revenues by filling in rear courtyards with additional buildings. Land values rose to reflect these returns. Assessments increased from twelve cents a square foot in 1807 to $5.71 a square foot fifty years later.[44]

By the third decade of the twentieth century, low-income housing was the characteristic form on the Lower East Side landscape. Seizing on this indigenous type, reformers asserted that the poor would (and should) not filter out of the community. Instead, it was argued that their neighborhood should be improved for them. A letter from a 1930s realtor and housing reformer James Felt reflected this attitude: "The theory that the lower East Side will become the center of some high class development may be sound, but after some 30 years experience first as a lawyer representing real estate investors and for 18 years in real estate itself, I think this is a wild dream. . . . I believe the East Side will remain what it is for many years to come, and the best that can come to it would be decent [low-income] housing."[45]

Roadway and Low-Income Housing, 1934. In 1934 the housing stock of the Lower East Side was decrepit, land values depressed, and population down. The opening of the outer boroughs and further suburbs with construction of new highways, bridges, and subways had escalated the population decline begun in the teens, when war and immigration restrictions diminished the flow of new residents to the district. In 1910, the eleven blocks between Grand, Scammel, Gouverneur streets and the East River, much of which would become the area's first publicly subsidized housing project, were home for twenty thousand people. By 1934, the population dropped to six thousand.[46]

Despite population declines, overcrowding persisted. Densities on the Lower East Side were 177 persons per acre.[47] In fact, this was the second densest community in the city, surpassed only by Harlem. As substandard buildings in the city were abandoned and demolished, and the depression progressed, overcrowding was exacerbated by people returning to the Lower East Side in search of cheap rents.[48]

Assessed values, which were eight dollars in 1930, were dropping. A 1934 comparison with two other areas shows just how low this property was valued. At Riverside Drive between West 94th and 95th streets, where removal of Dock Department commercial facilities was underway, unencumbered land was valued at twenty-seven dollars a square foot. At East 44th Street above Tudor City (in the vicinity of the United Nations), where the Swift Company then maintained slaughterhouses, property was assessed at $8.50 a square foot. On block number 263 between Cherry and Monroe Street, opposite Corlears Hook Park, the range was four to seven dollars a square foot.[49] While this valuation translated into fewer tax dollars, it also affected potential acquisition costs. The question remained: were these prices sufficiently depressed to allow low-income housing, or would developers seize the bargain and speculate that a higher residential use might be developed?

A turning point came in 1934, when a new source of funding for highway and low-cost housing appeared. The Public

Works Authority (PWA), created under the New Deal, provided federal money for low-income residential developments, jobs, and/or the reconstruction of slums. In July 1935 the city's new mayor, Fiorello LaGuardia, applied to the PWA for funds to complete the first section of the East River Drive. He reasoned that the roadway would induce "large scale improvements in the tenement house area to the west."[50]

Six years of fruitless discussion and elaborate plans were suddenly put aside; work on the drive was begun within a month. On 21 August 1935 the city took title to the entire East River waterfront from Corlears Hook north to East 14th Street, prompted by deadlines in spending the federal funds and the hurried need to provide jobs during the country's worst depression. A week later, the first tenement standing in the line of the drive was torn down.[51]

The potential for federal funds for low-income housing to replace the tenements in the line of the drive remained just that—potential—for three more years. Subsidies to buy land were a necessity. Yet ironically, advent of the drive raised land values well above the $1.50-per-square-foot loan eligibility limit set by Federal Housing Administrator Nathan Straus. Legislation was required in Albany before the newly created New York City Housing Authority could subsidize its own project.[52]

Now that the adjacent housing appeared to be slated for a less affluent group, the roadway took back its mantle of traffic relief. For the first time the East River Drive was described in news reports as a "continuous lane for fast traffic." In much the same language used in the 1920s for the West Side Highway, its immediate practical reason was to help move people quickly in and out of Manhattan and away from Park Avenue.[53]

The wish for grand housing, however, continued to be

strong. In an article entitled "What Do We Have to Look Forward to on The Lower East Side?" Chamber of Commerce leader Orin Lester wrote that "some properties should be improved for low-income housing. But the real effort should be redevelopment as a community to accommodate a residential and business population that can support the area."[54] Seven months later a drawing in the *New York Times,* accompanying the announcement that work would begin on the East River Drive, showed a wide boulevard lined on each side by parkland. To the west, on property now rezoned as residential, was an elegant array of art deco apartment houses.[55]

Housing the Poor, 1935–1938. Construction of the drive had begun, but the physical and social character of the Lower East Side continued to decline. Dorothy Rosenman, a public housing advocate who lobbied for redevelopment of the Corlears Hook area on behalf of the Henry Street Settlement (the neighborhood's chief social service agency), described the slum, "in a stone's throw of Wall Street and the Municipal Building, where thousands of people go to work each day, I came upon the most desolate, dilapidated scene. Homes closed down, houses that should be closed down. A picture of desolation and waste."[56]

Aged structures and numerous vacant and boarded-up buildings were a very large part of the problem. In the Corlears Hook district, a Lower East Side Chamber of Commerce survey disclosed people living in 942 old-law tenements, eighty-three new-law tenements, and 109 newer dwellings. In addition, here were seventy-one vacant lots, 120 boarded-up buildings, thirty-two structures closed or boarded-up above the ground floor, and eleven tenements in which only caretakers were to be found.[57]

This physical deterioration, combined with the devastating economic effects of the depression, caused social troubles to increase. Over 8 percent of the area's residents were classified as unemployable. Of the families who could work, the number on relief was high. In the Henry Street district, according to a Visiting Nurse Service report, there were many residents who were poorly nourished. With limited food and no heat and hot water in the tenements, grippe often became pneumonia. The mortality rate in this Health Center district was over 4 percent higher than for the city as a whole. Here too were more automobile accidents, more crimes, and more juvenile delinquents.[58]

The housing situation was also exacerbated by the very laws that had been created to improve living conditions. The Tenement House Department under Mayor LaGuardia had, in the mid-1930s, begun rigorous enforcement of the Multiple Dwelling Laws. Buildings deemed "unsafe and unfit for habitation" were razed simply because they were slums—not to provide room for a new structure. At the same time the depression had reduced housing starts. According to one housing analyst, five apartments were demolished for every two that were opened. Further, many banks and insurance companies had boarded up structures where mortgages had lapsed without being summoned by the legal process to do so. The receivers simply feared criminal liability should anyone be injured in these substandard tenements.[59]

Clearance and abandonment logically should have caused land values to decrease. At the close of 1936, housing reformers Dorothy Rosenman and James Felt, working with the Henry Street Settlement House, set out to prove that prices on the Lower East Side were sufficiently depressed to support new low-income housing. A land cost survey of eleven blocks to the south and west of Corlears Hook Park showed that contiguous parcels could be acquired at current assessment rates for a maximum of five dollars a square foot. This was $2.50 less than the land was valued at the beginning of the depression; yet even at this price, a subsidy would be needed to house low-income tenants.[60]

With this information in hand, Rosenman and Felt tried to interest private groups and the New York City Housing Authority in developing properties on the Lower East Side for public housing, and the first step was to pry loose the land from owners who were still waiting for a windfall. A survey of ownership patterns on the blocks between Jackson and Gouverneur, Water and Madison streets at the beginning of 1937 revealed that savings banks had substantial holdings there. The question was how to convince these institutions to turn over that land (which had minimal, if any, rental return and a high potential for liability) to the city for low-yielding Housing Authority Bonds. Felt and Rosenman proposed an incentive for the sale of these properties that capitalized on inflated assessment rates. The sale, they recommended, should be at current assessed values rather than at market rates, which were 50 to 75 percent lower.[61]

While negotiations were under way with the banks and the Housing Authority in preparation for a publicly supported project, officials at Henry Street also sought private funding. The purpose was to shelter low-income neighbors as well as to support the settlement programs. In 1938 board member Dorothy Rosenman and director Helen Hall approached the Carnegie Foundation for a grant. They envisioned a private demonstration housing project containing 650 dwelling units to rehouse local low-income families. There would also be space for Henry Street to build a cultural, recreational, and civic center with a full range of settlement house programs under the project roof. Kindergartens, health, play, and art facilities could all teach "neighborliness which steps into citizenship."[62]

The text of this proposal turned the East River Drive, opened 29 June 1937 from Grand to 12th Street, and the waterfront location into residential opportunities. The drive, they surmised, provided the potential to rehouse current residents of the Lower East Side. "Care has been taken to select sites which, because of vacant spaces and old and low structures, would call for least demolition. And at the same time would make sure that the housing project would face on the new drive, and parkway." And in wording that was added in the second draft, "and that its inhabitants would share in this new casement window the city is throwing open to the East Side."[63]

The timing of the Henry Street proposal to the Carnegie Foundation coincided with new changes in the roadway design. On 1 January 1938, Stanley Isaacs replaced Samuel Levy as Manhattan Borough President. Two and one-half months later Isaacs announced his "major undertaking" for the next four years: "to make the East River Drive a streamline highway . . . a continuous drive that will take care of both through traffic and business traffic that wants to travel long distances and relieve streets and avenues of their present burden."[64] The highway's purpose was no longer to serve the adjacent residential community.

Isaacs' announcement was prompted by the congestion on the portion of the roadway that was already open from Grand to East 14th streets. Described as "more like Riverside Drive" in layout, at least one lane of this section was already clogged with parked cars. Even the lights that had been installed to allow pedestrian access to the new East River Park impeded north-south movement. If, as forecast, the drive fostered adjacent development of any type, connections with local streets would increase local usage and further hamper through traffic.[65]

One can see that congestion on the new East Side route would rapidly be aggravated by noting the increase in motor vehicle registrations of the period. In the decade from 1926 to 1936, the number of cars registered in the city rose 53 percent and the number of people per car in New York City decreased from 11.2 to 8.5. In Manhattan, during the depression years when fewer people in the United States were buying cars, this ratio dropped from 9.0 to 8.3.[66]

Radical changes in motor vehicle and roadway design also helped remove all vestiges of an elegant boulevard from the newest proposal for the drive. The maximum rate that cars could travel in 1927 was around forty-seven miles an hour. By the date of Isaacs' announcement, those in a hurry could reach speeds of seventy-two or even eighty miles an hour. With faster cars, freeways called *autostrade* were opening in Italy, and the *Reichsautobahnen* were under construction in Germany. Keeping pace with this newest technology in New York State, in 1937 Governor Herbert Lehman signed legislation allowing the state and localities to establish such freeways.[67]

Political change and economic development are usually facilitated when principal actors are personally and socially connected. The lives of Dorothy Rosenman and James Felt were woven into such an intricate web of contacts. Felt was a lawyer and realtor in Lower East Side properties. Rosenman was married to an advisor to President Franklin Roosevelt who had been instrumental in New Deal legislation that funneled job and construction funds into the Lower East Side.

Earlier, as state senators, Samuel I. Rosenman, her husband, and Nathan Straus had worked together on the 1920s anti-nuisance laws for Riverside Park. Straus was now Federal Housing Administrator, setting limits on property values for federal loan eligibility. On her own, Rosenman cochaired the Housing Committee of an umbrella organization of New York City settlements, the United Neighborhood Houses. Associated with Rosenman in settlement work were Felt and Stanley Isaacs, Manhattan Borough President, in charge of completing the East River Drive.

A New Profile Emerges, 1939–1940. By the end of 1938, the pieces of the land acquisition puzzle for housing were in place. With compromises worked out between the city and the Federal Housing Administrator, the Corlears Hook area provided an opportunity for public intervention. With the Felt/ Rosenman survey as groundwork, Felt purchased options on a variety of lots at market rate. Then, on 18 July 1938, Borough President Isaacs wrote to Dorothy Rosenman, then vacationing in the Adirondacks, that "apparently the Housing Authority is taking over the area which you investigated so carefully and probably the very options which you started to obtain."[68]

Less than four months later Stanley Isaacs formally announced the subsidized housing project to be constructed opposite Corlears Hook Park between Gouverneur, South, and Madison streets. The project, to be called Vladeck Houses after a founding member of the City Housing Authority, would have apartments for 1771 low-income families paying rentals of $6.55 a room. Thirteen old-law tenements, eight new-law tenements (all built before 1905), Pechter's Bakery, garages, a car-wrecking yard and repair shop, and a trucking warehouse were all slated for removal. By April 1939 the East River Drive was once again in the news, and the arterial highway was still tied financially to the housing. The Montgomery to Grand Street section was approved, timed as a cost-saving measure to coincide with condemnation proceedings for Vladeck Houses.[69]

Public housing and an express highway were finally in place, yet proponents of higher-class housing remained. The Lower East Side Chamber of Commerce bemoaned, "all hopes for medium priced housing at Corlears Hook were shattered." A group of eleven savings banks tried in vain for a year to revive Rutgers Town. Robert W. Aldrich Roger charged that the city had "stolen" the site the banks had had in mind for white-collar housing since 1933. And a letter to the *New York Times* kept alive the concept of luxury housing. According to the writer, the Lower East Side could be "the most beautiful urban residential community in the world," where there was the opportunity for "comprehensive community planning . . . one great pre-conceived landscaped architectural composition . . . for the exclusive realm of the well-to-do."[70]

The wish for another kind of shelter remained just that. The cornerstone for the Vladeck Houses was laid on 3 October 1939, and the combined city and federal project opened six months later. From a letter from Helen Hall to Gerard Swope of the City Housing Authority, it was evident that the feat had been accomplished with style. "The people at Vladeck Houses," wrote Hall, "are so enthusiastic and excited that I often wish you could get the first hand comments that we do. One of the new tenants exclaimed the other day, 'You know I gotta paradise—I gotta four and a half room paradise!' "[71] It was paradise compared to what the private market offered low-income New Yorkers and like residents of other cities like Cleveland, Boston, and Buffalo that established housing authorities in the 1930s.

A simple ceremony, the removal of some barricades, opened the East River Drive from Montgomery to East 30th Street in May 1940. At a luncheon afterward, appropriately with the supporters of the Vladeck project, the highway was praised as having done its job. It would not only move traffic, it had helped to revive a neighborhood. In his concluding remarks, Chamber of Commerce vice-president Maurice Simmons spoke to its larger effects on the city: "The completion of the East River Drive and its ultimate connection with the West Side Highway and the Triborough Bridge will enclose our Borough in a riparian necklace which will adequately set off the beauty of the City's greatest physical asset, its unrivaled shore front. . . . "[72]

The only mention of the waterfront in the news was an article describing the view from a shabby Corlears Hook Park, which, now divided in two by the drive, was yet to be landscaped. Curiously, the activity that could be seen had once flourished on Manhattan's Lower East Side, but on this day it was hidden under mounds of landfill, a highway, and the first of many super-block public housing projects. Across the East River in Brooklyn was the New York Navy Yard.[73]

NOTES

1. *Comprehensive Plan for San Diego, California* (San Diego: Justin R. Hartzog Associates, 1926), 26; Thomas S. Hines, *Burnham of Chicago* (Chicago: University of Chicago Press, 1979), 182; and Condit, *Chicago 1930–70*, 332.
2. "The Plan of New York," A Letter to Frederick A. Delano, 24 November 1921, in *Committee for Plan for New York and Its Environs, 1923*, unpaginated.
3. City of New York, Department of Parks, *Annual Report*, 1900, 8;

Parks, *Annual Report,* 1898, 17–18; "The New Speedway on the Banks of the Harlem River in New York City," *Scientific American* 70 (March 1894):191, 199; and "The New Harlem River Speedway, New York" *Scientific American* 71 (April 1894):255, 263.

4. Henry Moscow, *The Street Book. An Encyclopedia of Manhattan's Street Names and Their Origins* (New York: Hagstrom Company, 1978), 57; and *Real Estate Record and Builders' Guide,* 25 December 1909, 1179.

5. *RRBG,* 5 March 1904, 482.

6. Ibid., 16 October 1909, 677; and Ibid., 15 August 1908, 332.

7. Ibid., 5 March 1904, 482.

8. Nelson Lewis, "City Planning," *City Planning Pamphlets,* n.p., 1915, 21.

9. New York City Improvement Commission, Report to the Honorable George B. McClellan, Mayor . . . (New York: Kalkhoff, 1907), 7–8; *RRBG,* 9 November 1907, 740; and Ibid., 16 November 1907, 788.

10. Ernest Poole, *The Harbor* (New York: Macmillan, 1915), 156.

11. N.Y. N.J. Port and Harbor Development Commission, *Joint Report,* 214–21, contains a list and descriptions of these proposals.

12. N.Y. N.J., *Joint Report,* 222–43: "Manhattan West Side Terminal Plans," Calvin Tompkins, *Report No. 2A* and *Report No. 6,* 1911; The City of New York, Department of Docks and Ferries, *Annual Report for the Year Ending December 31, 1910,* 7–8; *RRBG,* 9 September 1911, 346; and *RRBG,* 23 March 1912, 578.

13. "Drastic Changes in Traffic Facilities Will Alone Meet Manhattan's Needs," *The Port of New York* (January 1927):10–11; "Enlarge Manhattan Nine Squares by Extending It Southward," *The Port of New York,* undated, Mayors Papers, File Walker, Box 271, "City Planning," New York City Municipal Archives; *RRBG,* 24 February 1912, 390; and *RRBG,* 4 May 1912, 911. An even more elaborate scheme, involving a two-story structure housing open galleries overlooking the river and a roof-top highway with parkland on top of adjacent pier sheds, was proposed in 1898 by architect Alfred H. Thorpe, *New York Times,* 14 November 1898, 3.

14. New York City, Board of Estimate and Apportionment, Office of the Chief Engineer, *Report No. 32726,* 18 May 1926, 1–2, WL Papers, File "1926," The New-York Historical Society.

15. *RRBG,* 11 December 1915, 979–80; and Raymond J. Harrington, "The Elevated Public Highway Along the Hudson River Waterfront in the Borough of Manhattan," *Municipal Engineers Journal* 120 (26 November 1934):128.

16. *NYT,* Editorial, 16 June 1926, 24; and New York State, Ch. 484 L. 1926.

17. 8 April 1927, Mayors Papers, File Walker, "Elevated Express Highway."

18. Chief Engineer, *Report No. 32726*, 2.

19. Daniel Brogan, "Re-Planning the Lower East Side," *East Side Chamber News* 3 (April 1930):15. (Hereafter abbreviated *ESC News*.)

20. *NYT*, 21 August 1935, 21.

21. Borough President of Manhattan, *Annual Report, 1929*, viii, 3, 19; and *RRBG*, 2 February 1929, 5–6. The words "parkway, highway, and thoroughfare" are used interchangeably in this chapter.

22. City of New York, Department of Docks, Annual Report, 1935 (unpublished manuscript); DD, Cross Reference Sheets and Letters, 1920–40; and Federal Writers' Project of the Works Progress Administration in New York City, *New York City Guide* (New York: Random House, 1939).

23. Borough President of Manhattan *Annual Report 1929*, 15–16. (Hereafter abbreviated BPM.) Although this did not work downtown, Stanley Isaacs would use a similar tactic to reduce acquisition costs for the upper portion of the drive. In return for property improvements done by the city, private owners agreed to sell their water rights or rights-of-way to the city for one dollar, Stanley Isaacs Papers, File "East River Drive," Manuscript Collection, New York Public Library. (Hereafter abbreviated SI Papers.)

24. Letter from Stanley Isaacs to Howard Cullman, 15 January 1931, SI Papers, File Box 1, 1919–36, "Housing Assn. of CNY." In the three hundred blocks between the Brooklyn Bridge and East 14th Street, Fourth Avenue and the river, there were over six thousand owners, Joseph Platzker, "Land and Management on the Lower East Side," *ESC News* 4 (September 1931):14. The five waterfront owners were: Debby Ann Fay (4th to 5th Street), The Webb Institute (6th to 7th Street), Arthur Brisbane (7th to 8th Street), the Roach Estate (8th to 10th Street), Consolidated Gas Co. and New York Edison (10th to 14th Street), *NYT*, 19 December 1932, 3; and *NYT*, 25 August 1935, II:1.

25. Joseph Platzker, "Corlears Hook District (the Old 7th Ward)," *ESC News* 8 (May 1935):6.

26. BPM, *Annual Report 1929*, viii.

27. *NYT*, 24 September 1930, 25.

28. *RRBG*, 2 February 1929, 6.

29. Julius Malich, "Opposition to an East River Drive Parkway Is Negligible," *ESC News* 2 (May 1929):4; and, Malich, "500,000 Signatures Expected on East River Drive Petitions," *ESC News* 2 (June 1929):6.

30. New York City, Board of Estimate and Apportionment, *Minutes*, 7

November 1930, 7963–65. There had been periodic advisory planning commissions that served at the pleasure of the mayor. In 1928 the Merchants Association and Edward M. Bassett, father of the city's zoning resolution, convinced Mayor Walker to create the post of full-time, paid City Planning Commissioner. This position was authorized by the Sinking Fund on 10 October 1930, and Sullivan was hired at an annual salary of seventeen thousand dollars.

31. *The Regional Plan of New York and Its Environs,* 8 vols. (New York: Regional Plan of New York and Its Environs, 1929), Reprint (New York: Arno Press, 1974). One of the active members of this group was Harold Lewis, who as Engineer for the Board of Estimate and Apportionment was advisor to the 1904 City Improvement Commission that recommended the Chelsea pier design.

32. *The Regional Plan of N.Y. and Its Environs* (Arno ed.), 2:348–92.

33. Regional Plan Association, Press Release, 2 March 1930, 4. (Hereafter abbreviated RPA.)

34. RPA, Press Release, 1 July 1930, 5.

35. Ibid., 5–6.

36. "Putting a White Collar on the East Side," *The Survey* 66 (1 March 1931):585, 588–89. Miss Lasker's statements were made even more strongly by town planning consultant and professor at the Columbia University School of Architecture, Carol Arnovici. In an article entitled, "New York's Problem of Housing the Worker," *RRBG,* 26 March 1932, 5, he wrote, "It would seem to me logical that we should quit trying to rehabilitate Lower Manhattan for the poor and give it back to the well-to-do by building expensive, luxurious and well planned apartment houses in which the rich could live close to the financial district." Another approach to luxury housing on the Lower East Side waterfront may be seen in the futuristic drawings of Hugh Ferris, who in 1929 envisioned fifty- to sixty-story buildings hung from the cables of the city's bridges (three of which were on the Lower East Side). The luxuries included: landings for a launch, yacht or hydroplane with elevators to private apartments. "New York of the Future," *Creative Art* 9 (August 1931):161.

37. John Taylor Boyd Jr. and Arthur C. Holden and Associates, "Architects' Report," *ESC News* 3 (September 1930):9–10.

38. BPM, *Annual Report 1931,* 4, 22, 25; "Dawn on the East Side," *The Survey* 67 (15 December 1931):319; New York City Board of Estimate and Apportionment, *Minutes,* 30 October 1931, 7614–17; *NYT,* 21 February 1931, 19; Ibid., 20 March 1931, 47; Ibid., 14 September 1931, 3; Ibid., 19 December 1931, 36; and "Parks and Parkway in Shorefront Plan," *ESC News* 4 (September 1931):1.

39. *RRBG,* 6 May 1933, 3–5; BPM, *Annual Report 1934,* 7; City of New York, Office of the Mayor, "Descriptive Features East River Drive Grand to East 60th Street Borough of Manhattan, 1936"; *NYT,* 19 November 1933, 10:1; and *RRBG,* 31 August 1935, 4.
40. Gustave Zismer, "Thomas Proposes Apartments for 135 Blocks," *ESC News* 5 (May 1933):9–10.
41. Platzker, *ESC News* 8 (May 1935):6; Platzker, "East River District," *ESC News* 8 (July 1935):6.
42. "Rutgers Town Its Purpose and Principles," New York City Housing Authority Papers, File "Vladeck Houses," LaGuardia Community College Archives. (Hereafter abbreviated HA Papers.)
43. Platzker, *ESC News* 8 (September 1935):10; and Jackson, *A Place Called Home,* Ch. 13.
44. Jackson, 4; and Henry Street Studies, *A Dutchman's Farm: 301 Years at Corlears Hook* (New York: Henry Street Studies, 1939), 4, 6. (Hereafter abbreviated Henry Street.) John Jacob Astor, developer of worker housing in Chelsea, also owned property between Gouverneur and Jackson streets.
45. Letter to Dorothy Rosenman, 19 May 1937, Dorothy Rosenman Papers, File "1937," Manuscript Collection, Columbia University.
46. Henry Street, 12; and RPA, *Information Bulletin,* 26 February 1934, 1–2.
47. This information is extrapolated from acreage figures cited in RPA, *Information Bulletin,* 20 April 1931.
48. HA Papers, File "Vladeck Houses, 1939."
49. Annual Record of Assessed Valuation of Real Estate, 1934, Block Nos.:1253, 1356.
50. *NYT,* 28 July 1935, 12:2.
51. *NYT,* 21 August 1935, 21; and Ibid., 25 August 1935, 11:1.
52. Letter from Alfred Rheinstein to Nathan Straus, 17 January 1939, HA Papers, File "USHA, Vladeck Houses 1939 #1; and *NYT,* Editorial, 29 August 1935, 20. There were additional explanations for the escalation in land values. The city might have raised them to offset the tax loss from a long-term exemption granted to the buildings. Letter to S. I. from Fred M. French, 4 October 1934, File SI Papers, Box 1, "General Correspondence Isaacs." *RRBG,* 16 September 1939, 3, cites another reason: the Multiple Dwelling Law and a five-year tax exemption for improved buildings resulted in repairs and modernization that would have affected assessments.
53. *NYT,* 1 September 1935, 4:10.
54. *ESC News* 8 (February 1935):6.
55. *NYT,* 1 September 1935, 4:10.

56. Dorothy Rosenman, Radio Talk "NYU Hour," 15 December 1937, SI Papers, File Box 2, 1937–9, "Rosenman."

57. Platzker, *ESC News* 8 (May 1935):11.

58. Henry Street, 18; Reports from Visiting Nurses, Henry Street Settlement Collection, File "folder 102:12," Social Work History Archives, University of Minnesota, (hereafter abbreviated Henry Street Collection); and "Application for a Subsidy Contract by the New York City Housing Authority to the City of New York for a Low Rent Housing and Slum Clearance Project Known as Vladeck City Houses," 1 May 1939, 10, HA Papers, File Box N–Z, 1939, "Vladeck Houses."

59. Jackson, *A Place Called Home,* 209; and *NYT,* 28 July 1935, 20.

60. Letter to DR from James Felt, 24 November 1936, Henry Street Collection, File "folder 64:4."

61. Rosenman Papers, File "1937 and 1938." The city was allegedly making assessments at 128 percent of true value in order to keep its debt limit high to underwrite spending. Letter to Helen Hall from Assemblyman William S. Bennet, 4 August 1938, Henry Street Collection, File "folder 101:14." On 17 March 1937, the New York City Housing Authority formed a committee to look into pooling the savings bank properties to sell to the Authority, Rosenman Papers, File "1937"; and telephone interview with DR, May 1983.

62. Memorandum I, "Six Factors Which Would Distinguish the Henry Street Housing Project," 5, Henry Street Collection, File "folder 61:4"; and Minutes of staff meeting, 18 April 1939, Henry Street Collection, File "folder 67:3," 7–8.

63. Memorandum II, 2, Henry Street Collection, File "folder 61:4." The final version of these memos was printed as *A Dutchman's Farm: 301 Years at Corlears Hook.*

64. "Problems of the Office of the Borough President," Memo to the Citizens Union, 18 March 1938, SI Papers, File Box 2, "September 1938"; and SI to Walter D. Binger, Commissioner of Public Works, Draft Memo, 3 January 1938, SI Papers, File Box 3, "1938 East River and Harlem River Drive."

65. SI Papers, File Box 3, "1938 Press Releases"; and SI interview with Amie Brun, WHN, 21 February 1940, SI Papers, File Box 5, "East River Drive."

66. RPA, *Information Bulletin,* 21 March 1938, 4; and "Significant Trends in Highway Traffic in the New York Region," *Information Bulletin,* 12 December 1932.

67. Sam Bass Warner, *A History of the American City* (New York: Harper and Row, 1972), 41.

68. SI Papers, File Box 2, "1938 R"; and Memo from H. J. Morris to

Alfred Rheinstein, 28 November 1938, HA Papers, General File 1938, "Corlears Hook Project, 1938." Felt was offered 2 percent of the condemnation price for the properties on which he had purchased options. He subsequently donated this back to the Authority.

69. Walter Binger to SI, Memo, 5 April 1939, SI Papers, File Box 3, "East River Drive January–May 1939"; New York City Housing Authority, *Fifth Annual Report, 1939;* G. W. Bromley, *Landbook 1930, 1934, 1940,* Block Nos.:243, 260, 263–64; and *ESC News* 11 (July 1938):Cover, 1.

70. *NYT,* 12 December 1938, 18; "City Picks Corlears Park Area for First Municipal Housing Project With Occupancy Tax Funds," *ESC News* 11 (July 1938):1; Memo from Savings Banks Interested in the Lower East Side, 20 July 1938, Mayors Papers, File LaGuardia, Box 761, "Housing Projects Lower East Side"; and *RRBG,* 8 April 1939, 3.

71. 14 November 1940, Henry Street Collection, File Folder "86:3."

72. "East River Drive Boulevard Opened on June 29th," *ESC News* 10 (June 1937):9.

73. "East River Drive's Latest Link Will Be Open to Traffic Today," *New York Herald Tribune,* 17 May 1940, 21.

Part Three

So the ocean and the sky and the rivers hold the city in their grip, even while the people, like busy ants in the cracks and crevices, are unconscious of these more primal presences.

Louis Mumford, *City Development,* 1945.

CHAPTER 7

Elevating the Wall

The Bricks Loosen and Fall, 1940–1980. The basic pattern for
the configuration of Manhattan's waterfront seemed to be set
by 1940, and in the ensuing decades spaces were filled in
largely by expanding the activities already in place. More
windows were thrown open on the East River as a line of
public housing projects, aided by Title I of the 1949 Housing
Act, solidified along the East River Drive from the Brooklyn
Bridge to Stuyvesant Town at East 14th Street, and from
East 102nd Street to Highbridge. Private high-rise apartments
rose on the empty spaces among the hospitals and power
stations between East 23rd and 96th streets. In the early 1950s
the United Nations building replaced abandoned stockyards
from East 47th to East 48th Street. It housed a new international activity, and its accompanying park, concealing a
street-level portion of the East River Drive, gave a limited

public an opportunity to be near the water without visual interruptions.

The circle of Manhattan's arterial highways began to close with completion of the upper portion of the East River Drive in 1940. Within ten years, if traffic allowed, cars were able to speed in a continuous circuit around the outermost edge of the island. Even the old Harlem River Speedway had been lengthened as far north as Dyckman Street and had been renamed the Harlem River Drive. Here at the northern tip of the island, was the only remaining opening in the "riparian necklace."

Completion of East River Park in the 1940s added another link to the growing number of waterfront parks in Manhattan. (Map 13) Riverside, Inwood Hill, Fort Washington, and Highbridge parks formed a horseshoe–shaped green, albeit inside of the highways, around the top of the island from West 72nd to East 162nd Street (with the exception of twenty blocks between Columbia University's Baker Field and East 201st Street). Then Robert Moses and Stanley Isaacs eked out some riverside green from their highway projects for Harlem and the Upper East Side. An esplanade was constructed from East 125th Street to the East 60s, and the delightful new Carl Shurz Park was cantilevered out over the East River Drive where it ran at grade in the East 80s. Downtown, several years later, Robert Moses also expanded Battery Park.

The original layer of Manhattan's concrete rim cracked in the 1960s. The island's shipping activities, which had begun their disappearance in the late 1930s with a decline in local passenger ferry traffic and railroad lighterage, received a major blow from containerization. Huge, deep-draft container ships replaced smaller cargo vessels, and mechanisms capable of loading and unloading full railroad cars and the trailers of large trucks made the traditional manual, parcel-by-parcel

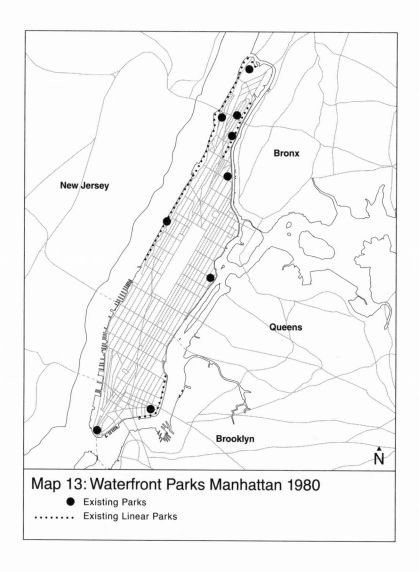

Map 13: Waterfront Parks Manhattan 1980

● Existing Parks
.... Existing Linear Parks

method known as break–bulk packing obsolete. Given a deep channel; large pieces of cheap, open land for container parking, pick–up and delivery; and a long, parallel docking space unloading and loading, which formerly had taken twelve days, could be accomplished in under twenty-four hours.¹ Thus, as in cities like Boston, Seattle, and San Francisco, Manhattan's many street-end finger piers were suddenly obsolete; and with the opening of the Port Authority's container port at Port Newark-Elizabeth, the move of shipping to New Jersey, so feared at the turn of the century, became a reality.

Harbor passengers, while still contributing to a substantial part of New York's economy, adopted a new mode of travel in the 1960s. Based on waters away from Manhattan, where land was expansive and cheap, jet planes from Kennedy Airport on Jamaica Bay in Queens lured travelers away from the time-consuming transatlantic luxury liner. Jets even captured the city's popular cruise trade with flights to southern ports where passengers could board ship in a warm climate. As the city planned for a new passenger liner terminal in the mid-1960s, providing eleven-hundred-foot slips for the *Queen Elizabeth* and the *France,* the number of transatlantic passengers through the Port of New York was in the midst of a decline from seven hundred thousand in 1955 to an eventual forty-two thousand in 1978. The need for still another group of West Side piers had already dissipated.²

Legally, in the 1960s the city's waterfront was still reserved solely "to aid navigation and commerce," and as Manhattan's percentage of the total domestic and foreign tonnage handled in the port began its decline from 19 percent in 1958 to 3 percent in 1971, the city still sought ways to revive its pier system. On the East River, where thirty-one of the thirty-five existing piers were over forty years old, and one-third were vacant or unusable, the city spent $7.3 million to build a new break-bulk pier. It immediately fell into disuse. Nearly half of the seventy-nine piers on the Hudson were in poor or very poor condition. To rectify this, another eleven million was expended on Pier 76; it was quickly transformed into the city's towaway pound for illegally parked cars. Modernization of the Chelsea piers, this time from West 14th to 37th Street, was the biggest disappointment and financial disaster. According to one report, within weeks of the opening of this $34 million project the shipping company that had leased the space decided that paying double rent—at Chelsea and at a container terminal elsewhere—would be cheaper than staying at the slow-moving, breakbulk Manhattan piers.[1]

Anticipating the container revolution, the Department of Marine and Aviation (the name assigned to the Dock Department beginning in 1942) also began to look for innovative ways to reuse the city's outdated and decaying pier system that did not rely solely on maritime commerce. In 1962, the Department had published a glossy volume that considered several new appendages to the island of Manhattan. A state-of-the-art convention center with bus terminal and parking garage, heliport, observation tower, and slips for sightseeing vessels and excursion craft was depicted on underwater land from West 38th to 43rd streets. Billed as an "unprecedented new city area," sixty-five acres of artificial land between Battery Park and Chambers Street were also shown, complete with hotel, yacht basin, office building, and thirteen "executive" apartment towers, each twenty stories high. Along the river, cargo sheds and six wharves would allow shipping to continue undisturbed.[4]

These visions quickly grew grander. In 1965, consultants for the Department of City Planning's *Lower Manhattan Plan* eyed the potential of the remaining underwater land between the bulkhead and pier-ends surrounding downtown Manhattan. Battery Park City and Manhattan Landing, the latter on 110 acres of decking over the piers between the Governor's Island Ferry Terminal and the Manhattan Bridge, were among the concepts under consideration.[5]

A redesigned West Side Highway was also included. Even as the highway was being opened in 1931, the Regional Plan Association had anticipated its demise. "If we look far enough ahead," the planners wrote, "it is not improbable that the [elevated West Side Highway] will be removed . . . that it will be possible to obtain some finer treatment of the whole West Side area." Now, just three decades later, the same organization talked of depressing this highway under excavations from the World Trade Center and of creating a continuous waterfront esplanade to take advantage of the great beauty of the downtown waterfront.[6]

The search for new productive uses of what had once been waterfront cities' principal asset was also on across North America in the 1960s, and direction for change came from several disparate sources. Large sums of federal money were available for urban renewal. This allowed Boston and Baltimore to clear collapsing sheds and piers, to install new streets and sewers, and to make impressive plans.[7] Sights were initially set on widespread clearance and the erection of high-rise residential and office buildings. Then, in 1966, the National Historic Preservation Act prompted concern for rehabilitation of cities' maritime treasures and supported an ethic for smaller-scale development.

The layers of brick, iron, and concrete that had amassed on urban waterfronts over the centuries limited who could be there. Now there were government attempts to reverse that. In 1965, the White House Conference on Natural Beauty urged that water courses be planned "for their protection and development to enhance human life and the quality of man's environment."[8] The Federal Water Pollution Control Act supported this sentiment, providing guidelines and incentives for cleaning up the nation's waterways. A year later in New York, in language strongly reminiscent of an earlier century, Manhattan Borough President Percy Sutton introduced a conference on the future of Manhattan's waterfront. Much of the shore, Sutton explained, "has fallen into decay and disuse despite the development of several modern piers. Suffering from years of neglect and divided authority, it has too long been regarded as marginal land, a dumping ground for industries, highways, rotting piers, and raw sewage. The public has been and continues to be denied access to the waterfront."[9]

California was the first state formally to acknowledge the opportunity that outmoded pier systems presented for people to use and enjoy the water's edge. Public assembly and water-oriented recreation were among the activities this state recognized as "essential to the public welfare" of the San Francisco Bay area as it was improved, developed, and preserved. Thus the term "public access" became current; eleven years later, in the federal Coastal Zone Management Act, it was made national policy.[10]

Spurred by government incentives, world's fairs, and a widespread availability of vacant property on downtown waterfronts, cities and private developers actually began to make piecemeal changes. Trade marts were begun, through partnerships between the city and private developers, on the downtown harbors of Baltimore and New Orleans. These modern towers were an effort to uplift the cities' declining maritime economies and the sagging images of their ports. In Seattle, tourism and recreation gained a foothold on the Elliott Bay waterfront in 1962. To accommodate the wave of visitors expected for the fair that summer, a private developer opened an inn on a pier formerly used for maritime

purposes, and mothballed ships were even turned into floating hotels and restaurants. Six years later, with the model of
Montreal's Expo '67 before it, the government of Toronto
began planning for Ontario Place, a new exhibition and
amusement park on unused landfill near to the declining
downtown harbor. Then in the early 1970s, the urban waterfront version of the suburban shopping mall appeared. An
entrepreneur opened San Francisco's Pier 39, and the Rouse
Corporation began work on Boston's Quincy Market.[11]

Meanwhile, with 75 percent of the East River shore already
devoted to nonmaritime uses, and with a scarcity of inland
development sites, investors once again began to acknowledge the presence of the rivers in Manhattan for their non-
industrial value.[12] The apartment towers of Waterside,
between East 25th and 30th streets, rose on decking in the
East River where coal barges once found meager landings.
Although it has been criticized for its starkness and exclusiv-
ity, the project was the first in Manhattan to combine housing
with direct access to the water's edge—without interference
from a roadway. Further uptown, Roosevelt Island was partially converted from an isolation ward to a riverside community. The publicly subsidized Riverbend, for moderate-
income tenants, and the AFL–CIO-sponsored 1199 Plaza
reoriented residents to the water with terraces and parks
that faced towards, rather than away from, the Harlem
River.

Then an event occurred that provided the extra impetus
needed to translate dozens of new waterfront visions into
reality—the nation's bicentennial. The stench that still per-
vaded rivers and harbors was forgotten in 1976 as millions
of people crowded at the Battery in New York, on a wharf
at Baltimore's spare Inner Harbor, and even at a former

steamboat landing in St. Louis. Entrepreneurs and preservationists, tugboat captains and office workers, children from land-locked ghettos and from luxury apartments gravitated to their primordial home—the water. Lining the shore, they watched parades of tall ships, boarded antique barks, and swarmed rivers and bays in pleasure craft of every shape and size.

Although few of the individuals who were caught up in this excitement would ultimately change their city's waterfront, the larger psychological effect was dramatic. People now believed they had a precious historic resource to preserve and that it was a resource which also could improve the quality of their lives. Citizens of waterfront cities across the United States suddenly awoke to the possibilities of recapturing their ugly, dirty shore.

Even when maritime activities were flourishing in New York, there were appeals for considering other functions for the waterfront. On 15 October 1893, the *New York Sun* had commented: "unlike most other capitals, New York sits directly on the seashore, the foremost maritime city in the world. There are a hundred miles of coastline. Commerce has the first claim in those hundred miles of water front, but no one will dispute that some of it should be saved for the pleasure and refreshment of the people." Three years after the bicentennial, with the ideological momentum that had been gathered, and with a more mundane need to improve the environment in order to keep business and residents in the city, New York's Mayor Edward I. Koch expanded on the *Sun's* sentiment. On 18 January 1979, in a speech marking the fortieth anniversary of the New York City Planning Commission, Koch announced: "if there is one thing I want my administration to be identified with, it is that we brought

the harbor back to the City of New York, that we built on
our greatest treasure, that we opened the waters to the people
of the city."

Rebuilding the Wall, 1980–1990. "Massive," "over-reaching,"
"untenable," "over-scaled," "embodiment of developmental
excess," and "bad urban planning," were some of the phrases
used by the press and the public to describe what began to
happen on Manhattan's waterfront after Mayor Koch made his
grand pronouncement.[13] As the International Longshoremen's
grip on its large inventory of decaying piers finally loosened,
the city geared up to revitalize the shore. In a booming economic climate, the aim was to derive the highest possible income from this land. Developers, aided by deep-pocketed
backers and armed with well-known architects, clamored at
the doors of City Hall and the Department of Ports and Terminals (the latest name for the Dock Department) to help the
city realize its goal.[14]

The era of waterfront mega-plans was at its height throughout North America in the 1980s. In Seattle and Toronto, Van-
couver and Baltimore, piers were lengthened for shopping
instead of shipping. Industrial buildings once used to store
bolts of fabric waiting for export, or to transform coal arriving
by water into lighting for downtown streets, were turned into
apartments, offices, museums, and amusement parks. And entire new communities were planned for abandoned railroad
rights-of-way.

Although the uses of waterfront space and facilities may
have changed, development patterns remained the same.
Rather than preserving the few remaining natural coves for
aquatic activities, or reutilizing surviving finger piers for fishing, floating pools, sunning, or simply staring, developers still

sought to extend the land from the water's edge to provide space for enlarged containerports or apartment building foundations. In Manhattan alone, there were plans to establish approximately 150 acres of new residential and commercial shoreline, an addition equal to 75 percent of the total new land that the city had been allowed to create under the Montgomerie Charter more than 250 years earlier. (Table 3; Map 14)

Similarly the pattern of massing layers of concrete one upon the other, and thus of separating city residents from their rivers, continued unabated. Of course, this practice was not confined to New York. As one critic wrote, "along the lakefront in Toronto . . . housing is edging right up to the water. At Union Wharf and the Charlestown Navy Yard in Boston, and in a number of small cities the same relationship of buildings to water prevails." [15]

A New Blueprint for the West Side, 1970–1974. Manhattan's waterfront has always had its visionaries. In the late nineteenth century, George McClellan imagined the shoreline encircled by working piers. His idea was endorsed by business and government and, over time, much of his plan was realized. In the first half of the twentieth century, Robert Moses dreamed of waterfront parks and circumambient highways, and he used his innumerable powers to carry out his dream. The Regional Plan Association and Edward J. Logue, president of the New York State Urban Development Corporation (UDC), constituted the next generation of idealists. Their vision for a comprehensive development of the West Side waterfront was rejected, however, and it became the basis for a pattern that consisted of planning hubris met by community opposition.

The scheme to build a new West Side Highway, from the Battery to 42nd Street, began inauspiciously. Its roots were in

Table 3. Landfill Proposed for Manhattan but Not Implemented: 1930–1985

Date Proposed	Legislation	Boundary	Location	Acres
1931–84	Lincoln West, private development	Bulkhead and 4 piers	West 59th to 72nd Street	14.3
1966–85	Special Manhattan Landing Development District (East River Landing)	Bulkhead to pierhead line	Ferry terminal to Manhattan Bridge	110 landfill deck
1968–	Riverwalk, Department of Ports and Terminals request for proposals	Bulkhead to pierhead line	East 14th to 23rd Street	24 deck
1972–85	Westway, Federal Highway Act	Bulkhead to pierhead line	Harrison to West 36th Street	181 landfill
1983–	Hudson River Center, Department of Ports and Terminals request for proposals	Bulkhead to pierhead line	West 36th to 40th Street	20 deck
	Total proposed landfill/deck			349.3

Sources: Richard Baiter, Office of Lower Manhattan Development, *Lower Manhattan Waterfront,* June 1975, 11; Development Plans for Lincoln West and Riverwalk; New York City, Department of Ports and Terminals; and Westway Management Group, *Summary of the Westway Plan,* n.d.

the UDC's 1971 *Wateredge Development Study.*[16] Logue and his staff (including a young man by the name of Richard Kahan, who would resurface in other major waterfront projects) saw an opportunity in the deterioration of the southern section of the Miller (or West Side) Highway, now more than forty years

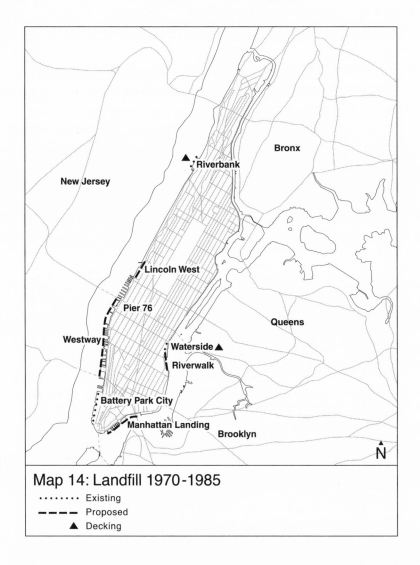

Map 14: Landfill 1970-1985

········· Existing
— — — Proposed
▲ Decking

old. They were also motivated by the availability of federal highway funds for road improvements. They envisioned a massive, 700-acre development on decking built out over the Hudson River, with traffic and subways below.

The rebuilt highway would be the planning mechanism to tie together the disparate West Side waterfront revitalization schemes under discussion since 1962. Battery Park City lay at the southern end of the route, the interpier areas no longer reserved solely for commerce, and its ninety-two acres of newly made land already in place. A convention center and superliner terminal were planned for the middle section, and a housing and athletic field complex was projected for the decking over the West 60th Street rail yards at the north.

Like the planners on the Lower East Side in the 1930s, supporters of the 1971 UDC proposal hoped that the highway project would also help to solve citywide problems, particularly a dearth of housing and recreation space. The land created on the decking would have accommodated 75,000 to 85,000 new apartments and would have yielded 150 acres of continuous waterfront park, connecting Robert Moses' Riverside extension to Battery Park.

The UDC plan sparked the imagination of government officials and spurred them to immediate action. By the end of 1971, the West Side Highway up to 59th Street had been designated an interstate, which made it eligible for federal funds. Governor Nelson A. Rockefeller and Mayor John V. Lindsay then created a steering committee for a "West Side Highway Project." [17]

The appointment of a committee satisfied federal funding guidelines, but two years later there was still no formal plan. The need to do something about the existing West Side Highway became acute when a portion of the elevated road suddenly collapsed, causing the death of a driver whose car fell through the crevice and onto the street below. The highway was immediately closed to cars from the Battery to West 44th Street. Local streets became heavily congested.

In April 1974, a new blueprint for the West Side Highway Project was released. [18] Westway (a name coined by Deputy Mayor John E. Zuccotti) was less ambitious than the UDC proposal, but it was still a significant undertaking. As described a decade later in the *New Yorker,* "virtually everything about Westway is enormous: its size (two hundred and forty-two acres of landfill; a six-lane roadway extending more than four miles . . .), its cost (officially two and a quarter billion dollars) . . . and its advertised benefits (sites for thousands of new housing units along the river; eighteen thousand man-years of new, on-site jobs; ninety-three acres of . . . parkland on the landfill, above an underground portion of the highway)." The landfill alone evoked gargantuan comparisons: "Extending up to three blocks out into the Hudson, it would require enough dirt to cover all of Central Park to a depth of six feet." [19]

Westway had the support of a host of government officials and prominent New Yorkers. Its guiding principles for West Side development spanned the administrations of three governors and three mayors, all of whom supported the project once in office. [20] Four presidents held office during the Westway period, and they took their cues from city and state officials. President Reagan issued the first federal highway check for the project. Senators Moynihan and D'Amato and former Mayor Robert F. Wagner Jr. were on board, as were David Rockefeller and two organizations that he chaired, the New York City Partnership and the Downtown Lower Manhattan Association. Scenting new jobs in the air, both organized labor and the New York Chamber of Commerce and Industry firmly backed the proposal. A specifically pro-highway group, Citizens for Balanced Transportation, was formed (boasting former Deputy Mayor John Zuccotti as its spokesperson), and there was even an intrepid group calling itself Westsiders for Westway. [21]

In addition to setting forth the traditional arguments that the project would provide a means to repair the highway at no cost and to increase the tax base, proponents of Westway, like supporters of the 1971 UDC proposal before them, saw in the venture a panacea for many of the city's ills. Like East River Drive advocates of the 1930s, they also believed the highway would bring businesses from the suburbs to the declining downtown financial center. The project meant jobs for an area with high unemployment, and the roadway would relieve exasperating neighborhood traffic conditions.

City officials facing the possibility of bankruptcy also saw Westway as a relatively cost-free means to recapture the waterfront and to rebuild the deteriorating West Side, though few saw it, as Logue had, as an opportunity to plan comprehensively for the resurgence of the West Side. Development continued in an ad hoc manner, with most parties assuming that Westway would be built by hook or by crook. Construction began on Gateway Plaza, the first of the Battery Park City residential units, and on the convention center. Plans for the railroad yards, however, had been dropped due to a cutback in federal housing funds.

A Waterfront in Limbo, 1974–1985. More than money would be needed to realize Westway. Just as the Women's League for the Protection of Riverside Park had fought to save the waterfront from the railroad in 1916, public interest groups now organized to defeat the highway. Their numbers were large and their tactics, drawn from those used by similar efforts to fight the completion of the Embarcadero in San Francisco and Route I-95 in Baltimore, were far more sophisticated than the Riverside Park efforts.

Opposition to Westway centered on several issues. There were strong parochial interests. New Jersey officials feared that the landfill on the New York City side of the Hudson would kill the market for new housing on their shore. Meanwhile, in the absence of a land-use or zoning plan for the landfill, both Community Board 2 and the West Side Ad Hoc Committee Against the Interstate were concerned that some ill-defined development would block off the waterfront with high-rise structures, increase the population to undesirable levels, and bring in even more traffic.

Board 2, representing the historic Greenwich Village area, made an effort to analyze the proposal but could not agree on an alternative. Using funds from the West Side Highway Project, they hired preservation architect John Belle to help them understand the transportation technicalities of the Westway Environmental Impact Statement (EIS). Belle found merit in their fears. As the highway would have limited access and egress, local truck traffic probably would increase. Belle's recommendation to the Board, which was never circulated, was to substitute for Westway a street-level boulevard on the existing West Street. Parkland could be constructed along the Hudson, and minor development could occur on the piers.[22]

Robert Moses also opposed Westway. Self-interest may have played a part: for the first time in decades he had been left out of a major public works project. In reaction, he put forth a scheme entitled "West Side Fiasco," in which he described an alternative roadway extending up to his Riverside Park improvement. Between 72nd Street (where the park ended and the elevated Miller Highway ascended) and 59th Street, Moses proposed that the elevated structure be moved inland and placed at the eastern end of what were then active rail freight yards.[23] The topography of the land at this location was such that the roadway could be either depressed or at grade, leaving room for new housing over the yards and a magnificent waterfront park.

There also were groups with broader interests. In the latter years of Westway, the National Taxpayers Union seized upon the Reagan Administration's growing fiscal conservatism. They argued that the estimated $2 billion needed to build Westway would worsen the $200 billion federal deficit.[24]

Environmental issues were another rallying point. Groups such as the New York City Clean Air Campaign (whose president, Marci Benstock, would become renowned for her single-minded devotion to fighting Westway) and the Sierra Club may have had to scramble for money, but they were not at a loss for members. The Natural Resources Defense Council and environmental lawyer Al Butzel and his partner Mitch Bernard provided pro bono legal assistance. They were supported by the American Rivers Conservation Council, the Appalachian Mountain Club, the Hudson River Sloop Clearwater (and its director, folk singer Pete Seeger), Friends of the Earth, and the Hudson River Fisherman's Association. Additional traffic, they all alleged, would increase air pollution, and the dredging operations for the landfill could pollute the water, destroy tidal wetlands, and disturb the habitat of the fish that had only fairly recently returned to the cleaner waters of the Hudson.

Another group of Westway adversaries latched onto the issue of mass transit. In 1973, under pressure from a growing number of anti-highway groups around the country, Congress had passed legislation enabling localities to trade in any money designated for unbuilt sections of the interstate highway system in their districts. In return, an equivalent amount of federal money could be requested to be used for mass transit and a simpler roadway. Groups including Action for Rational Transit, Business for Mass Transit, the Clean Air Campaign, and the City Club pressed for the trade-in. They expected Congress to appropriate the entire $2 billion Westway price-tag for bus and subway improvements and repairs on West Street/Twelfth Avenue.

To fight the highway, the anti-Westway assemblage employed many of the maneuvers that the Women's League had found useful in combatting the railroad earlier in the century. The protesters met with the leaders of the Westway project and with environmental agencies. They wrote letters and editorials, and testified at public hearings. They even convinced the federal Environmental Protection Agency (EPA) to hire a member of a protest group to analyze state studies.

At various stages in the design process, an anti-highway contingent of New Jersey legislators used their votes in Congress to hold up funds requested by project leaders. They even garnered enough votes to deny an extension of the 30 September 1985 trade-in deadline. This tactic bolstered the spirits of mass transit advocates and became a crucial element in Westway's ultimate demise.

The formulation of an alternative highway plan was another device that helped to bring an end to Westway. Despite the notable negativism and diverse agendas of the opposition, member groups shared one common objective: they wanted to stop the project. They could not agree, however, on what should take its place. The result was a waterfront in limbo—decaying, unusable by the public, and providing no tax base for the city. Esteemed environmentalist René Dubos described it aptly in a *New York Times* op ed piece: "New York City has been blessed by nature with waterfronts much more spectacular and richly diversified than those in any other large city in the world, but it has made a mess of this natural endowment."[25]

In August 1980, Paul Willen, an architect who (with his partners at Gruzen Samton) had been working on the West

60th Street railroad yards and other waterfront redevelopment projects since the 1960s, teamed up with John Belle to bring to light Belle's earlier boulevard plan. River Road, as it was called, was a manageable vision of how the trade-in funds could be used for highway and waterfront renewal. The proposal called for a modified roadway and waterfront park, both on existing land. A six-lane arterial would be constructed along West Street/Twelfth Avenue. At various points along its route the road would dip down below grade. In those sections, it would be covered by a deck that would form pedestrian promenades over the road, connecting the upland with the river and piers.

Here, for the first time, was a credible alternative to Westway that local residents, mass transit proponents, and environmental groups could rally around and that public officials could embrace. Instead of being dismissed as just another crazy move by the opposition, River Road was discussed seriously by Westway officials.[26] Even Mayor Koch, a Westway supporter since becoming mayor in 1978, embraced River Road. This more modest concept fell victim to a higher political agenda, however, and Koch soon reversed himself.[27] Still, although it was not accepted, the River Road proposal had served to open the eyes of Westway advocates to the possibilities of a cheaper and more environmentally acceptable solution.

Environmental legislation, such as the Federal Water Pollution Control Act of 1965 and the National Environmental Policy Act, allowed opponents to pursue a simple strategy that delayed Westway for eleven years, contributing to the demise of the project. At each milestone—the demolition of the elevated highway or the issuance of the water quality permit, for example—the groups filed suit. Initially these suits alleged that the state had failed to prove Westway's construction would not harm the environment. During the decade of delays, the growing cost of the highway, the availability of the federal trade-in moneys, and a surging interest in Lower West Side real estate development provided the grounds for additional litigation. However, these suits alone were not enough to defeat the perceived behemoth; all but one of the law suits were dismissed by the courts.

A passing comment made to Al Butzel, the opponents' counsel, would provide the issue that cracked the Westway case and allowed the opponents to prevail. In a last ditch attempt to convince the regional director of the EPA to veto the landfill permit that had finally been issued in 1980, Butzel, representing the Clean Air Campaign and the Sierra Club in litigation against the permit, traveled to Washington. Although he came home without a veto, he was asked whether he had seen the latest fish studies. During the federally required environmental hearings, the state had maintained that the areas between the piers that were slated for filling were a biological wasteland. In view of the pollution that had been accumulating for nearly a century, this position seemed intuitively correct. In fact, much to everyone's surprise, studies conducted by the state found thousands of fish flourishing there, among them striped bass, an edible species that had been in decline for years. A subsequent study confirmed the importance of this particular habitat to the bass, but it was repressed by Westway officials. The opposition finally had the issue that would make their court case strong enough to defeat Westway.[28]

"Somewhere That's Green": The Westway Park, 1978–1985. Although the Westway vision originally included a large park, without an assurance of landfill there had been no reason to plan for this amenity. Several years before the River Road idea

became public, however, a number of commentators and community activists, among them former Village resident Jane Jacobs, had proposed other solutions. They asserted that an adequate park could easily be built on the existing piers.

By 1978, the West Side Highway up to 26th Street had been closed for four years, and intrepid West Siders were finding new uses for it. In the West Village, developers were converting warehouses on Greenwich and Washington streets (just one to two blocks from the increasingly decaying riverfront) into rental and co-op apartments, and creating a demand for additional recreation space. People began to bicycle and jog on the abandoned roadbed of the elevated highway. It even became the focus for street artists, who used the expanse as an easel. (Figure 32)

Two years later, however, the commissioners of the state parks and transportation departments announced the award of a $2 million contract to architects Venturi & Rauch and landscape architects Clarke and Rapuano to design a ninety-three-acre Westway Park. The park was officially touted as a means to reclaim the waterfront, but the state had other motives, as René Dubos explained: "In addition to its obvious advantage for New Yorkers and tourists, Westway Park and other amenities along the New York waterfront will illustrate to the rest of the world that great cities can recapture scenic beauty and human quality even though more and more people move into them." [29]

The timing of the state announcement, however, made the park appear to be a palliative to head off protest against the increasingly controversial highway. In the same month, Carey and Koch signed a Memorandum of Understanding for Westway.

Then, in November 1980, demolition of the West Side Highway north of 26th Street began. By virtue of the island's

Figure 32. West Side Highway, 1978. (Courtesy of the author).

contour, a visitor standing as far away as Eighth Avenue could finally see clear through to the Hudson River.

Even as the state agencies announced the Westway Park design contract, there were rumblings of discontent. With federal dollars for the highway not always available when needed, critics asked, who would pay for construction of the park and for its eventual upkeep? Others pointed to the complicated design and questioned how construction of a park over a highway would be managed. Many felt that the state parks department might not be capable of getting the job done, and, furthermore, the parks and transportation agencies were thought to be working at cross purposes. Many faultfinders pushed for the creation of a separate entity, a special State Parks Commission for New York City that had authority over Westway Park was deemed a solution.

The Regional Plan Association had a more radical idea. Their *Annual Report for 1981–82* called for "a single West Side executive deputized by the governor and mayor to have full responsibility for the project including the park." Here were shades of Robert Moses, who, in describing his 1930s West Side Highway and park project a few years later, would comment that "only stubborn leadership availed, not the smoothness of diplomacy." [30]

Planning for the park proceeded slowly. The landfill issue was still embroiled in litigation, and there was an official need to play down the importance of the park in order to justify federal highway funding. Early in 1983, state parks and transportation officials unveiled not one, but three plans for Westway State Park. (Figure 33) The objective was to allow community members and citywide parks advocates to choose among the alternatives. Of course, it was also a way to create a constituency for speeding up the construction of the park, and thus the highway. If the public became excited about a park on this extraordinary waterfront site, the thinking went, they might fight for it. The landfill and highway would then be an easier sell.

But the public would not be fooled, and the number of Westway detractors grew. The Citizens for Balanced Trans-

Figure 33. Model of Westway State Park, 1983. (Courtesy New York State Office of Parks, Recreation and Historic Preservation).

portation, as well as members of a commission for the Westway Park appointed by Governor Carey and his successor, Mario M. Cuomo, worried about the failure of the city to plan the remainder of the landfill. Following the tenets of Jane Jacobs, they maintained that an urban park requires residences and commercial spaces on its borders to enhance safety and to enliven the open space.[31]

Environmental opponents had other concerns. Marci Ben-stock, John Mylod of the Hudson River Sloop Clearwater, and Bunny Gable (representing Friends of the Earth) focused on the shortage of money and the overlong timetable. "The park is a mirage unlikely ever to be funded and will not be available for 20 years," they wrote in a 1984 press release. Using the waterfronts of Boston and San Francisco as examples, they too echoed Jacobs when they recommended alternative parks that could be built within two years at a much lower cost. "Piers can be used for restaurants, marinas, recreation, ships and possibly housing," they suggested.[32]

One group, however, remained ardently supportive of Westway State Park. In a letter designed to boost the declining morale of its constituents and with little other than imagery to back it up, Westsiders for Westway wrote, "we've been waiting and dreaming of a park . . . you actually see the bikeway, sweeping lawns, the finger piers, even sailboats tacking in the wind."[33]

Ultimately the tenacity of the opposition, a serendipitous question about the striped bass, the rapidly approaching deadline for the trade-in, and the solid vision of an alternative roadway combined to deal the final blows to Westway. When, in August 1985, Judge Grisea revoked the permit to begin work on the Westway landfill, the trade-in benefit possibilities had been re-formulated by Congress to give the city $1.5 billion over five years. Although this value was less than many had anticipated, it was nonetheless hard to refuse after thirteen years of plans and studies but no action on the highway. Koch and Cuomo deliberated over whether to appeal the court's decision and to press for the now more than $2 billion waterfront development. But with the 30 September trade-in deadline and an unknown legal future, the officials decided to go for the money they knew they could count on.[34]

In the fourteen years since the publication of UDC's *Wateredge,* little had changed on the Westway route. A report by the West Side Task Force, appointed to recommend a new scenario for the West Side Highway and for the Hudson River waterfront, described what was to be found there: "West Street has been left as an 'interim roadway,' poorly surfaced, aligned and drained, with crossings which serve to discourage pedestrians from reaching the Hudson River. In the water there remains an irregular line of partially broken-down pier structures, some unused, some salvaged for recreational purposes and essential municipal or other public services, and some used for parking." [35]

Pie in the Sky, 1980–1990. Even though the failure of the Westway proposal left miles of unaltered shore, Manhattan's waterfront had begun to change in the first half of the 1980s. The mood in the city was expansive. Mayor Koch had announced his waterfront goal of building on this great treasure at a time when the city's fiscal situation had vastly improved, and the financial markets were on an upswing. (Figure 34) Government-backed waterfront projects were moving along at a fast pace. A master plan had been prepared for Battery Park City, and ground was broken for the esplanade, for parks, and for the World Financial Center. Shortly after this commercial complex opened and the Rector Place apartments were put on the market, an announcement proclaimed "Battery Park City Eyes Fifth Office Tower," and a memo circulated among city and state officials suggesting that the entire project be extended northward along the Hudson to West 34th Street. [36]

Elsewhere in Manhattan, the opening of the South Street Seaport was heralded as "the biggest" of the Rouse Company's gallery of "urban extravaganzas," with promise of surpassing

Figure 34. Floating Beach, 33rd Street and the East River, 17 August 1983. (Courtesy Corbis-Bettman).

Baltimore's Inner Harbor and Boston's Quincy Market. When the Javits Convention Center (between West 30th and 38th streets and Eleventh and Twelfth avenues) opened its doors, press descriptions told of an edifice so long that the Empire State Building could lie down in its main corridor, and so wide that two 747 jumbo jets could fit inside wing tip to wing tip.[37]

But even as the bulldozers began excavations and ribbons were cut, architectural critic Ada Louise Huxtable voiced disappointment with the direction of change: "I have very mixed feelings. . . , " she wrote, lamenting the Seaport's passing. "I guess that what I am really doing is saying goodbye. Because what will surely be lost is the spirit and identity of the area as it has existed over centuries—something that may only be important to those of us who have loved the small shabby streets and buildings redolent of time and fish, or shared the cold sunlight of a quiet Sunday morning on the waterfront with the Fulton market cats, when the 19th century still seemed very much alive." [38]

Despite such discontent from some quarters, attempts to redevelop Manhattan's waterfront continued unabated. The focus was on three major sites: the Upper West Side railroad yard area adjacent to Riverside Park, the East River bank above Corlears Hook, and New York Harbor.

The West 60th Street railroad yards had been owned by the Penn Central Railroad (a successor to the New York Central) until 1975, and thereafter had been sold to a succession of developers. Once a major transportation center, part of the yards had been covered by Robert Moses with decking to enlarge Riverside Park. There remained seventy-six acres ripe for development. (Figure 35)

The yards were in a massive pit, braided with unused railroad tracks. They were rimmed on one side by the island's rock formation and on the other by a continuation of the elevated Miller Highway that landed at grade at 72nd Street. This portion of the elevated road still functioned, effectively cutting off the waterfront with a four-lane structure that ranged from twenty to sixty feet high.

Nonetheless, because of its size and its accessibility to transportation, and because the waterfront was not highly valued as an amenity, there had been several proposals for the interior portions of the property beginning as early as the 1960s. Af-

Figure 35. Abandoned Penn-Central Railroad Yards, West 59th to 72nd Street, 1985. (Courtesy the Trump Organization).

fordable housing—and, as the economy and the neighborhood improved, luxury housing—appeared in prospectuses. Of course, industry still had a hold on the waterfront, and developers explored commercial options as well as residential alternatives. One enterpriser tried to interest the *New York Times* in moving its plant to the site.

The elevated roadway presented an obstacle to any development and kept designers from even contemplating a park on the waterfront. In one plan, the highway had actually been moved from the water, only to be replaced by buildings. Most plans included a token park located in the middle of the property, in the shadow of low- and high-rise structures.[39]

There were two city-owned waterfront properties of interest to real estate developers in the early 1980s. The area on the
East River between East 14th and 23rd streets, known in the
nineteenth century as Stuyvesant Cove, contained the only remaining large cove in Manhattan. The waterfront was mostly
abandoned, save for a cement batching plant and scattered
parking. This seemingly insignificant parcel was described by
architectural critic Paul Goldberger as "one of the great sites
in the city. The shore bends inward into a cove offering a
view back inwards towards the Manhattan skyline."[40] But
even though the cove was visible from the neighborhoods in
that area, the site was physically cut off from Manhattan by the
elevated FDR Drive (formerly the East River Drive), which
had been constructed in the 1940s. In the blocks to the west
lay the private enclaves of Stuyvesant Town and Peter Cooper
Village, the residents of which coveted their uninterrupted
views of the river and their parking spaces under the highway.

Like the rail yards, the cove had also been considered for
development prior to 1980, and the first scheme was as grand
as some of the plans put forth under the McClellan regime in
the nineteenth century. Paul Willen (the architect who had
designed River Road), together with his partners, had laid out
several of the early plans for the rail yards. For the cove, they
envisioned building cofferdams into the East River, pumping
out the water, and filling the underwater void with the requi-
site industrial buildings. On the roof, above water, would be
2,500 apartments with a park in the center.

Others had a different vision for this area. In the mid-1970s,
the city, recognizing a dearth of public open space in this
community, targeted the cove for a twenty-four-acre waterfront park. In fact, the waterfront was already effectively being
used for this purpose. Some residents actually crossed under
the highway and used the aging concrete bulkhead for sun-

ning. The more intrepid among them swam daily in the river, which, at times, looked remarkably clean.[41]

But by 1979, in the midst of a fiscal crisis, city officials began to consider the site's potential for higher economic return, and turned their backs on the park that had been envisioned for the cove as early as 1935. To stave off anticipated community protests, officials announced that the city would require the inclusion of a park in any plan put forth by private developers.

The city was correct in its assumption that the cove had monetary value. In 1979, Alfred Bloomingdale, president of Diners Club, and Melvin Stier, a suburban mall developer, offered a plan to Mayor Koch to build a commercial center on the six acres of existing land and on decking over the twenty-four acres of water. In response, claiming that waterfront property was inalienable and that city property could be leased only through a competitive process, the deputy mayor and the commissioner of the Department of Ports and Terminals (the agency that managed the property) announced the issuance of a Request for Proposals (RFP) for a mixed-use project for the East River cove that would be "the largest waterfront revitalization project to date." In order to conform to city policy, the waterfront was to be open to public use.[42]

The other city property up for grabs, known as South Ferry, lay between State and Broad streets at the tip of Manhattan, adjacent to Battery Park. It housed the outdated Whitehall ferry terminal and the deteriorating Battery Maritime Building, one of two remaining nineteenth-century pier structures in Manhattan. (Map 15)

Buoyed by the response to the East River site, the city departments of Transportation and Ports and Terminals soon issued an RFP for South Ferry. Critic Goldberger, in reviewing the ensuing proposals, was expansive in his description of "perhaps the most extraordinary building site in lower Manhattan." Another writer saw the property as "potentially a spectacular entry to downtown, as well as a magnificent termination to Manhattan, but at present . . . in sorry disarray. The river is nearby but mostly unseen. Ferries come and go, but invisibly; busses not boats dominate the space."[43]

There had been no prior interest in South Ferry. But city officials hoped that its physical prominence and a growing demand for downtown office and hotel space would inspire developers to vie to rehabilitate the decrepit ferry terminal and the neighboring white elephant. South Ferry held a high place on Koch's wish list of city-owned sites to be auctioned off.

The avid responses to the opportunities presented by the three waterfront development sites reflected both the bullish economic times and the absence of a coherent vision for the city's waterfront. Between 1981 and 1986, developers proposed a total of eight schemes for residential and commercial uses at the rail yards and the cove. There were eight commercial proposals alone for South Ferry. The place-names, unfortunately, lacked vision. South Ferry Plaza, East Cove, River Cove, and Riverwalk reflected the waterfront setting. Other names, such as Television City and Trump City, reflected the use the developer wished to draw to the site, or the developer's ego.

Prominent people led the development teams: Abraham Hirschfeld, owner and operator of parking garages and a perennial candidate for mayor and other city offices; Donald Trump, real estate mogul of Atlantic City casino fame; Daniel Rose and William Zeckendorf Jr., both from prominent New York City real estate families; John Zuccotti, the former deputy mayor who was active in Westway; and Richard Kahan, an author of *Wateredge* who went on to become president of the Urban Development Corporation and Battery

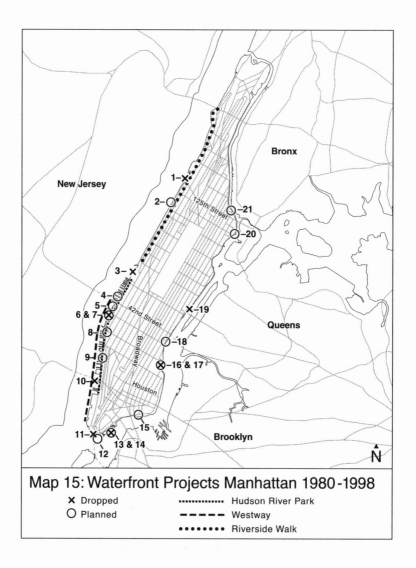

Map 15: Waterfront Projects Manhattan 1980-1998

✗ Dropped ············· Hudson River Park
○ Planned − − − − Westway
•••••••• Riverside Walk

KEY

1. Harlem-on-the-Hudson
   (1981–1993)
2. Riverside Walk (1991–)
3. Lincoln West/Television/
   Trump City (1981–1991)

4. Passenger Ship Terminal
   (1992–)
5. New York Waterways Ferry
   Terminal (1993–)

Park City. World-renowned architects constructed models and presented glossy pictures of the proposed development projects. Helmut Jahn and Alexander Cooper worked on the tip of the island as well as the railroad yards. Lewis Davis created designs for South Ferry and the cove. Architectonica, Fox and Fowle, Hardy Holzman Pfeiffer, Kohn Pederson Fox, Gruzen Samton, Ulrich Franzen, and I. M. Pei all had visions for changing the city's waterfront. (Figure 36)

Were all three of the sites to be developed, the waterfront would be radically transformed. But the change would be driven by finance, not by an overall plan for the waterfront. The city's open space inventory would increase, but it was the other proposed land uses that would pay for the parks, create the jobs, and bring in the rental income the city sought in return for leasing out its waterfront property. "Highest and best use" was the order of the day.

If there *was* a vision for the waterfront, it was in the architecture. The three-story Whitehall Ferry Terminal might be covered by a structure ranging from thirty-five to sixty stories in height; shaped as a box or perhaps a faux Eiffel Tower; or topped by a form resembling the Empire State Building, a

---

6. Hudson River Center (1981–1995)
7. Pier 76 Heliport (1996–   )
8. Chelsea Waterside Park (1986–   )
9. Hudson River Park (1986–   )
10. Westway (1971–1985)
11. South Ferry Plaza (1984–1991)
12. Whitehall Ferry Terminal (1991–   )
13. Manhattan/East River Landing (1966–1991)
14. Piers 9, 13-14 (1992–   )

15. Pier 35-6 Walkway/Community Center (1992–   )
16. Riverwalk (1979–1992)
17. Stuyvesant Cove (1990–   )
18. East 34th Street Ferry Dock (1991–   )
19. East 60th Street Heliport/Hotel (1981–91)
20. Washburn Wire Works (1986–   )
21. Harlem Beach Esplanade (1991–   )

Figure 36. Model of South Ferry, 1985, Architectonica, Architects. (Courtesy of the author).

glass conservatory, or "an outrageous beacon." The land uses envisioned by planners ranged from pedantic to entertaining: offices, hotel rooms, a public library, theaters, museums, an electronic people mover, or a Red Grooms sculpture.[44]

According to the various development plans, the predominate use for the cove and rail yards would be residential; the architecture would be tall and dense. At the East River cove, which was one-third the size of the yards and one-fifth the size of the developable land of Westway, entrepreneurs called for up to 1,876 apartments, or sixty-two apartments per acre. A hotel and assorted retail and commercial activities were also to be included. Proponents of the "East Cove" proposal envisioned four seventy-story towers, which would become the tallest residential structures in the city.

Donald Trump, who bought the rail yards in 1985, was inching up to the Westway planners in hubris. The preliminary scheme for the newly made land at the latter site had envisioned 140 dwellings per total developable acre; Trump sought 7,600 apartments, or 100 units per acre. This level of building would bring 20,000 new residents into the community. In addition, not to be outdone by the "East Cove" developers, Trump proposed the world's tallest edifice, standing forty-eight stories higher than the Empire State Building. (Figures 37, 38)

By 1986, Westway had been abandoned but not much had changed in the world of waterfront development. Trump had replaced Helmut Jahn with Alexander Cooper and was shepherding his Trump City through the lengthy city approval process. The world's tallest tower was still in the picture.[45] On the East Side, the city had selected Riverwalk, massive for its site, with five high-rise apartment structures (each from thirty to forty stories tall), an eighteen-story hotel, and a fifteen-story office tower. In addition to shops and a small marina, plans included eight acres of parkland, though not all of it would be publicly accessible. All of this would be constructed on platforms in the river. Paul Willen and his Gruzen partners were the initial architects. At the tip of Manhattan, William Zeckendorf Jr. led the consortium selected to develop South Ferry Plaza into an office tower that was expected to generate $375 million for the city over twenty years. Among his backers was the Continental Development Group, chaired by Richard Kahan.

Despite the optimistic scenarios, there were voices of caution. One commentator noted that before Donald Trump could build "either the world's tallest building or the rest of the

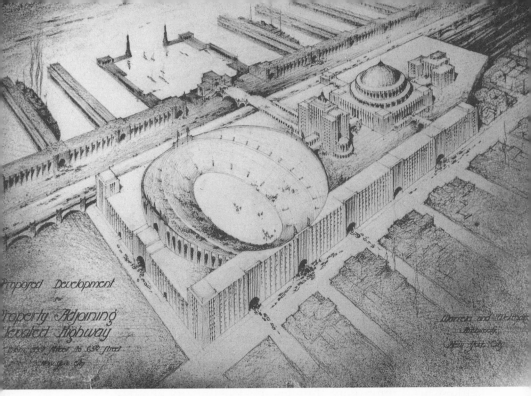

Figure 37. Proposed Development for the Railroad Yards, West 59th to 65th Street, Warren and Wetmore, Architects, July 1929. (Courtesy New York City Municipal Archives).

sprawling Trump City complex he envisions on the Upper West Side waterfront, he will have to work his way through a sometimes humbling city approval process that has led to the overhaul of many a sweeping plan." Reacting to the South Ferry project, another critic questioned whether the project was worth pursuing and remarked that "the city will soon be sponsoring an additional huge building in a part of town that is virtually suffocating from tall buildings already . . . it will close off a part of the island that has been open forever." [46]

Splinter Groups and Land-Use Brawls: The Opposition Continues, 1986–1990. The approval process for each of these projects was

236

Figure 38. Model for Television City, West 59th to 72nd Street, 1985. (Courtesy the Trump Organization).

daunting in itself, and any proposed use of the water only added to the problems. In 1930, the Regional Plan Association had synopsized the implementation problems it foresaw:

It is not within the power of any one body to carry into effect a plan for any portion of the waterfront of Manhattan. There are concerned in the control of this waterfront federal, state and municipal authorities. Numerous private corporations and persons are concerned in its ownership and development. The preparation of any plan with the expectation of public action in putting it into effect is impractical. The expense that would be involved even with the highest degree of cooperation between the owners and the public authorities would be enormous. The making of plans for definite

237

application has to proceed in the usual piecemeal way, although these partial plans should be fitted into a comprehensive plan of the city.[47]

A half century later, the land for Riverwalk, slated to become housing and recreation space, was industrially zoned. Yet once rezoning occurred, ferries would be precluded from docking there, as landings were permitted only in industrial areas. Should the South Ferry Plaza developers try to introduce a floating recreational structure into the harbor, it could land them in the same morass experienced by the owner of the River Café (a floating restaurant in Brooklyn) who wrestled for thirteen years with, among other obstacles, fire codes applicable only to land-based buildings. In addition, agency fragmentation and overlaps were legion. A former deputy mayor inventoried at least fifty federal, state, regional, city, borough, and community bodies with power over the shore.[48] And, of course, there was always the opportunity for an environmental lawsuit to delay things further.[49]

Unlike the top-down process that had defined Westway, this time around the city did attempt to head off some of the problems engendered by opposition to waterfront development. Officials solicited Community Board 6's comments when they reviewed the Riverwalk proposals. They also included potential adversaries in the panel that selected developers for South Ferry. They invited groups such as the Municipal Art Society, the Parks Council (a citywide advocacy group), and Community Board 1 to view models and to present written comments on the design. But, in the case of both Westway and Riverwalk, the city had set no standards for comparison and the reactions were splintered. In the final analysis, decisions were based not on a preferred design or on the waterfront's assets, but on economic benefit to the city.[50]

Figure 39. Model of Trump City, Opponents' Perspective, Undated. (Courtesy the Riverside South Planning Corporation).

Tempers flared rapidly once work on the Trump City and Riverwalk approvals began. Community Board 7 reviewed and rejected the scale, the density, the overcrowding of the subway, and the lack of amenities planned for Trump City.[51] (Figure 39) On the East Side, Community Board 6 commissioned students at the Harvard Graduate School of Design to study the district and to give it credible goals to fight for. Responding to the cries of Stuyvesant Town and Peter Cooper residents, the study recognized the needs for parking space and "to optimize views for the benefit of the local community."

Waterfront recreation, open space and continuous public access along the water were still high on the list.[52]

Flush with victory over Westway, opponents of the succeeding waterfront projects—backed by environmentalists and some elected officials—gathered together splinter groups to form People for Westpride, the Coalition for a Livable West Side, and Citizens United Against Riverwalk. A journalist using pugilistic imagery conjured up the earlier West Side battle. The Trump approvals, he wrote, would "set in motion the biggest land use brawl since Westway."[53]

People who had never been active in community affairs came to life as captains in every Peter Cooper building. They sent postcards, letters, and petitions to the Mayor, the City Council, the Borough President, and anyone else who would listen. They testified at every opportunity. Community Board 6 would avail itself of a new venue to voice opposition: the Uniform Land Use Review Procedure (ULURP).[54]

The two West Side groups had different agendas for battling Trump. The Coalition, consisting mainly of Lincoln Towers residents whose views would be totally blocked by the development, wanted to prevent the project altogether. Westpride, on the other hand, was willing to accept development at the lower scale that had actually been approved by the city for an earlier proposal.[55]

Despite their different goals, the two groups' tactics were similar in many ways. In addition to traditional lobbying, law suits on environmental grounds were a potent tool that both groups intended to use from the outset.[56] But unlike the hand-to-mouth existence and outsider role suffered by prior waterfront advocacy groups, Westpride held glamorous fund raising parties to help pay for consulting attorneys and engineers. With this expertise came drawings, statistics, and credibility. And, by working with city planning staff to come up with a mutually acceptable development solution, Westpride became the first community group privy to the dialogue between the city and the developer.[57]

At the end of the decade, the real estate market was reeling from the 1987 stock market crash. Nonetheless, the Zeckendorf group was working out complicated transportation alignments for cars and ferries at the South Ferry site, even though busses still dominated the area. Trump, still under strong fire from the community and unsure of his financing, continued to negotiate with the city for higher densities and bulk limits for Trump City. Then the Koch administration, as one of its last acts, certified Riverwalk to begin ULURP.

On 21 February 1990, in an overflowing auditorium at Washington Irving High School, an event was held that is still talked about in the community. A very carefully orchestrated hearing informed the city that Riverwalk should be turned down. In order to lend credence to the opposition, made up primarily of residents with parochial interests in maintaining views and parking, local Assemblyman Steven Sanders and the Chair of Community Board 6 seized on a familiar tool—the effects on the environment. Led by the Sierra Club and supported by good government groups that had been called in for the occasion, speakers at the hearing, one after the other, decried the pressures that would be inflicted on an antiquated and overtaxed sewer system, warned of the dangerous river pollution that would ensue, and condemned the platform with its pilings and shadows that would kill any fish habitat in the East River. The ULURP battle was enjoined.[58]

A few months later Paul Goldberger commented on the planning excesses of the 1980s: "There is much talk these days of vision, or the lack of it, and frequent questions as to just what vision in city planning should consist of. It's a fair thing to ask, since vision is often thought to mean impractical, far-out gestures, pie in the sky schemes that are the opposite of realistic."[59]

NOTES

1. For discussion of Manhattan's declining waterfront see Robert F. Wagner Jr., "New York City Waterfront: Changing Land Use and Projects for Redevelopment," *Urban Waterfront Lands* (Washington, D.C.: National Academy of Sciences, 1980), 78–99; and Mitchell Moss, "Staging a Renaissance on the Waterfront," *New York Affairs 6* (2 November 1980):3–19.

2. Moss, "Staging a Renaissance on the Waterfront," 9; Parsons, Brinkerhoff, Quade, and Douglas, *Development of the North River Waterfront North of 59th Street, Manhattan,* 12 December 1963; and New York City Department of Marine and Aviation, *Consolidated Passenger Ship Terminal,* April 1965.

3. The Port of New York Authority, *Pier Survey,* 1972, 7–8; Ebasco Services Inc.; Moran, Proctor, Mueser, and Ruttledge; and Eggers and Higgins, *The Port of New York Comprehensive Economic Study for Manhattan North River: Development Plan 1962 to 2000,* 28 November 1962, Plates 45–70; Regional Plan Association, *The Lower Hudson: A Report of the Second Regional Plan,* December 1966, 19–20; and City of New York, City Planning Commission, *The Waterfront: Supplement to Plan for New York City,* January 1971, 24. In 1970, the state legislature amended the reservation of the city's waterfront land to allow "any business, commercial, maritime or public purpose," Moss, "Staging a Renaissance on the Waterfront," 12.

4. Ebasco, 27–29; 13–16.

5. Wallace, McHarg, Roberts, and Todd; Whittlesey, Conklin, and Rossant; and Alan M. Voorhees and Associates, *The Lower Manhattan Plan,* 1966; and Richard Baiter, Office of Lower Manhattan Development, *Lower Manhattan Waterfront,* June 1975.

6. RPA, *Regional Plan of New York and Environs,* 2:389; and *The Lower Hudson,* 1966, 67.

7. Wrenn, *Urban Waterfront Development,* 124–32, 146–56; and Farrell, *Development and Regulation of the Urban Waterfront,* 3–19.

8. "Water and Waterfronts," *Summary Report,* 15; U.S. Congress, Senate, Water Quality Act of 1965, Pub. L. 89–234; and Water Quality Improvement Act of 1970, Pub. L. 91–224.

9. Introduction to conference brochure, 19 November 1966.

10. McAteer/Petris Act, 1965; and Amendments to the 1972 Coastal Zone Management Act.

11. See Brooks et al., "The Resurgence of Urban Waterfronts"; Farrell, Gemmil, *Ontario Place;* Hershman et al., *Seattle's Waterfront;* and Wrenn.

12. City of New York, Department of Marine and Aviation, *Redevelopment East River Piers Lower Manhattan,* 2 June 1959.

13. Paul Goldberger, "Another Chance for a Prime Piece of Real Estate," *NYT,* 1 July 1990, 25; Community Board 6, "Resolution Re: The Riverwalk Project," February 1990; and Joan Lebow, "Fate of Waterfront Plan in New York Seen as Barometer of New Mayor's Views," *Wall Street Journal,* Real Estate, 18 February 1990.

14. The Department of Ports and Terminals was the heir to the Department of Marine and Aviation. In 1986, it was renamed Ports, International Trade & Commerce. The agency was disbanded in 1991, and its properties and tasks were distributed among the Economic Development Corporation, the General Services Administration and the Buildings Department.

15. Craig Whitaker, "Rouse-ing up the Waterfront," *Architectural Record,* April 1986, 67.

16. Urban Development Corporation, *Wateredge Development Study, Hudson River Edge Development Proposal,* May 1971; see also RPA, *The Lower Hudson,* 34–35.

17. The city and state parks commissioners were both members of the steering committee. All of the West Side Community Boards were added between 1973 and 1979.

18. This was the "Outboard Alternative" published in the U.S. Dept. of Transportation, Federal Highway Administration, *West Side Highway Project: Draft Environmental Impact Statement,* 1974. (Hereinafter abbreviated as DEIS.)

19. "The Talk of the Town/Notes and Comment," 4 March 1985, 33; and Tom Morganthau et al., "The Death of a 'Boondoggle'?" *Newsweek,* 19 August 1985, 28.

20. Lindsay and his successor, Abraham Beame, were supportive from the outset. As a candidate, Edward Koch, a Greenwich Village resident, vowed that "Westway will not be built." Once in office, he became an ardent Westway champion, but he threatened to withdraw his support during election years. *New Yorker,* 4 March 1985, 34.

21. Letter from John J. Messinger, Business Manager of the International Union of Operating Engineers, to the President, 8 November 1983; New York Citizens for Balanced Transportation, *Newsletter,* 2 January 1984, 1; and New York State Office of Parks, Recreation and Historic

Preservation, File, "Westway, Public Participation." (Hereinafter abbreviated as OPRHP.)

22. Paul Willen, architect, interview by author, 24 October 1997.
23. Testimony of Robert Moses, consultant to the Triborough Bridge and Tunnel Authority, to hearings on DEIS, 25 November 1974. The EPA wrote off the idea in its response, alleging that the highway realignment would pose a negative impact on development and would necessitate an incursion into Riverside Park. Daniel Gutman, Environmental Planner, telephone interview by author, 1 December 1997.
24. Sam Roberts, "The Legacy of Westway: Lessons from Its Demise," *NYT,* 10 July 1985, A1.
25. René Dubos, "Restoring a Treasure," *NYT,* 23 July 1980.
26. Willen, interview.
27. For more than two years Governor Hugh L. Carey, a Westway supporter, had threatened a mass transit fare increase. Koch, running for a second term, could not afford the alienation that would be engendered by a fare increase. Carey agreed to keep the transit fare at an acceptable level and, in return, Koch gave up on River Road. Ironically, had River Road prevailed, the trade-in would have provided the funds to keep the fare from rising. Letter from Koch to Honorable Hugh Leo Carey, 16 December 1980, OPRHP, File, "Westway, City Position."
28. Al Butzel, former Westway attorney, interview by author, 1 December 1997. In order to minimize the implications of the striped bass discovery on the proposed landfill, the state conducted another study hoping to show that the fish were distributed up and down the Hudson and that the loss around Westway would be negligible. Again, they were surprised. The bass were clustered only around the Westway piers.
29. René Dubos, *NYT,* 23 July 1980.
30. Roberts, *NYT,* 10 July 1985.
31. The author was a member of the Governor's Commission.
32. 19 June 1984, OPRHP, File, "Westway, Public Participation."
33. Ibid.
34. James Gill, Koch law partner, conversation with author, 3 November 1997.
35. *Final Report,* 8 January 1987, 4.
36. *Crain's New York Business,* 24 February 1986.
37. *Crain's New York Business,* 31 March 1986, 8.
38. Ada Louise Huxtable, "Development at the Seaport," *NYT,* Op Ed, 25 February 1979.
39. Willen, interview. The 1929 *Regional Plan* recommended using air rights over the railroad yard for offices and apartments. Between 1962 and 1985, there were five proposals by private developers. The Amal-

gamated Lithographers Union wanted to use federal housing funds to build Litho-City over the working rail bed; 5,000 moderate income apartments were proposed, with the *New York Times* as the commercial tenant. The intent was also to extend Riverside Park and to create a marina along the shore. In the 1970s, the Educational Construction Fund considered building 12,000 moderate income apartments, and schools with adjoining athletic fields. Donald Trump purchased the yards in the mid-1970s and hired the firm of Gruzen Samton (Paul Willen, partner) to design a project with towers on the waterfront replacing the highway. The community negotiated Trump down from more than 12,000 to 4,000 moderate income apartments. Abraham Hirschfeld took an option on the property in 1981 and considered building an amusement park and casinos. A year later, an Argentine developer, Francisco Macri, purchased two-thirds of Hirschfeld's option to develop Lincoln West. This project was finally approved by the Board of Estimate in September 1982. There were to be 4,300 luxury apartments and a twenty-six-acre park. Three years later, delays, higher costs engendered by city approvals, and the failure of Macri's financing caused him to sell his stake to Donald Trump. His first plan was for Television City, with 7,600 luxury apartments, a commercial shopping mall with NBC as the anchor tenant, and a 150-story tower designed by Helmut Jahn. In addition, forty acres of open space would be located in the interior and a thirteen-block waterfront promenade situated to the west of the elevated highway. (The project was renamed Trump City when NBC backed out.) Frank Sommerfield, "76 Acres, River View," *7 Days,* 15 March 1989, 38–41; Riverside South promotional material.

40. Paul Goldberger, "Decision Nears on Gracing East River," *NYT,* 21 June 1980.
41. The author curated a waterfront exhibit at the Municipal Art Society in 1989. Among the photographs was an image of a woman in a bikini, standing wet in the East River cove. The water was so clear that both her feet and the bottom rocks were visible. When photographed, the woman boasted that she had been swimming here for over a decade.
42. Peter Newell, "Rivercrawl," *East Side Express,* 10 March 1983, 1.
43. "Reshaping South Ferry: 8 Plans Before the City," *NYT,* 12 June 1985, B1; and 1976 Consultant Report on Whitehall Ferry Terminal, "Design Guidelines for RFP," 22 February 1986, New York City Economic Development Corporation, File, "South Ferry." (Hereinafter abbreviated EDC.) EDC was the successor to the Public Development Corporation, which, in the early 1980s, had inherited all of the major

waterfront development projects from the Department of Ports and Trade.

44. *NYT,* 12 June 1985.

45. In order for Trump City to be certified, Trump had to provide the City Planning Commission with the proper environmental studies and other documents. Once certified, the project would enter the city's Uniform Land Use Review Procedure (ULURP).

46. Martin Gottlieb, "Trump's Plans for 150-Story Tower on West Side Face a Strenuous Review," *NYT,* 20 November 1985; and Goldberger, *NYT,* 12 June 1985.

47. Regional Plan Association, Press Release, 2 March 1930, 7.

48. Wagner, "New York City Waterfront," 96.

49. For this reason, Trump elected not to build anything into the Hudson River that required complex approvals.

50. The author participated in the South Ferry proposal review process as an officer of the Parks Council. The project then was in the hands of the New York City Public Development Corporation (PDC), which had taken over nonmaritime development properties from the Department of Ports and Terminals. PDC's selection criteria for South Ferry mentioned the waterfront and open space. The civic organizations tried unsuccessfully to assure that these criteria were given greater weight. As an example of just how splintered the local groups were, when the Waterfront Committee of Community Board 6 reviewed the Riverwalk plans, the vote was: East Cove, 3; Riverwalk, 2; River Cove, 2; East River Development Project, 0; and no development at all, 2. Carol Piper, District Manager, Community Board 6, telephone interview by author, 21 November 1997; Irwin L. Fisch, "Waterfront Committee Has Pei Day," *The Community Herald,* 13 June 1980, 1–2; and "How Not to Plan a City," *NYT,* Op Ed, 27 June 1980.

51. Mary Muska, Executive Director, Riverside South Planning Corporation, interview by author, 21 November 1997.

52. Harvard Graduate School of Design, "East River Development, East 14th–East 59th Street, Manhattan, NY," 1980; and Piper, interview.

53. Michael Moss, "Long Fight Looms over Trump Digs," *New York Newsday,* 25 June 1989, 5. On the East Side, Riverwalk would face a "fight some observers say could reach the intensity of Westway." *NYT,* 21 June 1980.

54. Jane Crotty, former Chair, Community Board 6, interview by author, 20 November 1997; and Carol Piper and Mary Muska, interviews. The 1976 and 1989 New York City Charter revisions gave Community Boards the power to hold public hearings and furnish an advisory vote

on any project in the community within sixty days after it has been certified by the City Planning Commission.

55. Lincoln West had been approved for 4,300 luxury apartments. Steven Robinson, Architect and founder of Westpride, telephone interview by author, 1 December 1997.

56. Gutman, interview.

57. Robinson, interview.

58. Joyce Young, "1000 Neighbors Hit Developer's Riverwalk Plan," *Manhattan,* 11 October 1987, 1; Piper, interview; Testimony, Community Board Six, Manhattan, File, "Riverwalk, 1990 Public Hearing."

59. Goldberger, *NYT,* 1 July 1990, 34.

CHAPTER 8

I Must Go Down to the Sea Again

By 1990, Manhattan's waterfront was in need of serious repair. The accumulated layers of concrete were in a state of disintegration due to age and neglect. The same toredo and marine borers that had threatened the nineteenth-century port were feasting on the wooden piers. The elevated highways that still stood were unsafe, and officials questioned the structural integrity of Valdeck Houses. Moreover, the waterfront parks had little connection with the rivers.[1] Fortunately the 1990s would be a decade for assessment, for planning, and for re-opening the waterfront in a more realistic way.

New York's experience was similar to that of aging shore-front cities all over North America during this period. To keep its patrons interested, Baltimore's Inner Harbor switched from candle shops to Barnes & Noble and Planet Hollywood. Leaders in cities that had not been "Rouse-ified" sought other ways to enhance their economies. Gambling barges and riverboats took their stations dockside from Biloxi, Mississippi, to Davenport, Iowa. Touted as water-dependent structures, most of the vessels never even left the shore; the riverfront was walled off by utility areas, parking lots, and huge casinos erected on the barges.

In cities with a heritage of waterfront planning, advocates had their work cut out for them. In Chicago they sought to convince officials of the need to save portions of the newly cleaned up Chicago River for the public. A small not-for-profit group called Friends of the Chicago River was created in response to a real estate boom in this forgotten part of the city. Their lobbying spurred officials to adopt open-space design guidelines for the first time. The results were attractive, well-lit public promenades and seating areas located on private property along the once highly polluted river.

In the past, both natural disasters and money for public works had helped to wall off waterfronts. Now they bared it. In 1991, the wrecking ball hit San Francisco's Embarcadero Freeway. This event took place just two years after a massive earthquake had peremptorily settled a divisive highway debate.[2] With an accessible waterfront, city fathers transformed Pier 7, the longest shipping pier in San Francisco Bay, into an award-winning recreation pier, now heavily used for fishing, sitting, and strolling.

On the east coast, Congressman and Massachusetts favorite son Tip O'Neill secured federal highway funds for the depression of Boston's Central Artery, long the barrier between the mixed-use shoreline and the inner city. The effects of the project reached all the way out into the harbor, as the spoil from dismantling the artery and tunneling a new roadway was used to create the base for a highly publicized waste treatment plant surrounded by a large harbor park.[3]

Manhattan's waterfront in 1990 was at once unchanged and in further decline. Los Angeles had replaced New York as the country's preeminent port.[4] The remaining vestiges of Manhattan's supremacy, the Passenger Terminal piers built on the Hudson between 48th and 54th streets during the Westway years for transatlantic travel, were being repositioned as a venue for cruise lines. Meanwhile, there were the occasional twenty-four-hour "cruises to nowhere," and prison barges, a new type of floating facility.[5]

Newly elected Mayor David N. Dinkins took the oath of office in 1990, when the city's real estate market was flat and waterfront development was at a standstill. Once-prized apartments at Battery Park City were up for auction, and the entire northern half of the site was barren, with no developer interest on the horizon. Two major projects were on hold: Hudson River Center, a four-block hotel complex planned during the expansive 1980s to stretch along the Hudson just south of the Javits Convention Center; and East River Landing, Manhattan Landing reduced to a high-rise office and residential project slated for twenty-three acres of decking over the East River adjacent to the Seaport. The developers of South Ferry Plaza would soon let their option lapse, Riverwalk had been nicknamed "Rivercrawl," and architect Peter Samton suggested that the "black hole of land" at the Trump yards "be landmarked as a site perpetually to be developed."[6] (Map 15)

Finally a Plan, 1990–1993. The Dinkins administration began its work on the waterfront knowing what advocates had declared for years: that without controls, the development that had been viewed as beneficial for the opening and regeneration of the waterfront in the 1980s now threatened to block off and privatize the shoreline. The newfound popularity of the waterfront in the 1980s had also created a tension between competing needs: for mixed-use development, for open space, for the remaining maritime and industrial uses, and for preserving a fragile ecosystem. Meanwhile, developers, unsure of the rules and still faced with a confusing and lengthy approval process, were finding it increasingly difficult to secure financing for their projects.

Back in 1871, business interests had pressed the city to make order out of chaos, and the result was a workable plan to build piers. In 1984, the call came from the Parks Council in its *Agenda for Action*. Echoing the voice of the Regional Plan Association in 1930, the *Agenda's* primary recommendation was that "New York City should adopt a waterfront land use plan for each of the five boroughs . . . and each borough study must also be developed within the context of the whole city."[7] Barbara Fife, one of the principals behind this report, was now a deputy mayor.[8]

With a friend in City Hall and a lull in the real estate market, planning for the next cycle of New York waterfront renewal gained momentum. Following the lead of Boston's 1984 *Harborpark* and Toronto's *Regeneration Report,* waterfront planning became the cannon of the 1990s.[9] New York City's Department of Planning, with the assistance of a thirty-five-member panel of agency representatives, elected officials, and civic leaders, began to attempt to reach consensus on the future of the more than 500 miles of city shoreline.[10] Concurrently Ruth Messinger, who had succeeded Dinkins as borough president, concentrated on a plan for the Manhattan waterfront.[11]

For political expedience, the plans included something for everyone. They codified waterfront land uses, zoning regulations, and open-space designs for future generations. They also, for the first time in the city's history, reversed the time-honored policy of filling in the shore during periods of growth. Henceforth, new piers and platforms could be constructed only for maritime or recreational use.[12]

On a smaller scale, other groups also began to make plans

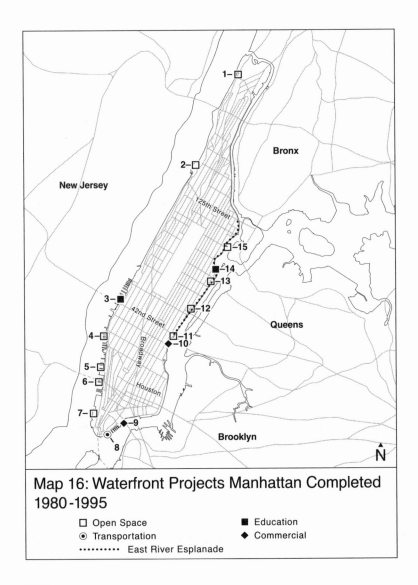

Map 16: Waterfront Projects Manhattan Completed 1980-1995

□ Open Space   ■ Education
⊙ Transportation   ◆ Commercial
·········· East River Esplanade

KEY

1. Dyckman Street Marina
   (1980–1994)
2. Riverbank State Park
   (1967–1993)

3. Intrepid (1980–82)
4. Chelsea Piers (1991–1995)
5. Morton Street Vent Shaft
   (1990–1991)

for the waterfront at this time. Battery Park City officials took advantage of the real estate surcease to re-examine its 1979 Master Plan and Design Guidelines, now more than a decade old. The new goals were to integrate the undeveloped, northern half of the site into the adjacent Tribeca community and to increase the marketability of the remaining parcels. (Map 16)

Immediately to the north of Battery Park City, the New York State Department of Transportation was refining plans for the Route 9A Reconstruction Project, which included a boulevard, promenade, and bikeway to be built from the Battery to West 59th Street using the Westway trade-in funds. There was also a plan for Route 9A's entire western edge. The West Side Waterfront Panel, made up of representatives appointed in 1988 by the state, city, community, and Manhattan borough president, published its concept plan for the design and financing of a new, 270-acre Hudson River Waterfront Park to be situated on the land, water, and piers (most of which was once to be Westway). After years of debate over the fate of the West Side waterfront, there finally was consensus on a park. It would be several years, however, before the plan would be released and funded.[13] (Figure 40)

Civic groups, which for years had protested waterfront projects only to find that the shore was now clearly worse off,

---

6. Pier 34 (1994–1996)
7. Battery Park City (1968–1996, Esplanade and waterfront parks complete, mixed use ongoing)
8. Wall Street Heliport (1983–1986)
9. South Street Seaport (1966–1985)
10. Waterclub Restaurant (1980–1982)
11. East 34th Street Esplanade and East River Esplanade Park (1989–1992)
12. East 60th Street Pier Park (1991–1995)
13. East River Esplanade (1984–1989)
14. Fireboat House Environmental Center (1980–1985)
15. 107th Street Recreation Pier (1986–1991)

Figure 40. Model of a Pier in Hudson River Park, 1997. (Courtesy Peter Rothschild, Quennell Rothschild Associates/Signe Nielsen).

sought a nonlitigious solution. The vehicle was planning. Daniel Gutman (once a consultant to the EPA on Westway), working for Westpride on a critique of Trump City, decided to formulate an alternative plan for the area north of 59th Street. Gutman drew on responses he had found in EPA highway files to Robert Moses' testimony against Westway, and on

RIVERSIDE PARK SOUTH
MUNICIPAL ARTS SOCIETY · REGIONAL PLAN ASSOC. · THE PARKS COUNCIL

Figure 41. Riverside South Park, The Civics' Alternative, 1990, Paul Willen, Architect. (Courtesy the Parks Council).

a rendering of the plan Paul Willen had prepared for Donald Trump in the mid-1970s. Gutman proposed to move the Miller Highway (the only remaining elevated portion of the West Side Highway) to the east. This would create a wide-open park on the river, with development as an extension of Riverside Drive.[14] (Figure 41)

The selling of the "civic alternative," as Gutman's plan was called, followed the usual path. For several years the idea lay dormant. Although Gutman worked for Westpride, there was dissension within the group over abandoning the tactics that were successfully holding up Trump City: working within the

255

system to thwart Trump's plan, and litigating when the opportunity arose. Finally Gutman, also a member of the Parks Council's Waterfront Committee, showed his scheme to Linda Davidoff, its Executive Director. Davidoff applied her advocacy skills to the task of marketing the alternative plan. Armed with engineering studies by Andrews and Clark that showed the feasibility of moving the highway (they had executed the anti-Westway design for Robert Moses) and a new drawing by Paul Willen, she gathered together a small group of influential civic organizations, including the Regional Plan Association and the Municipal Art Society, to lobby for the plan.[15]

Like River Road, the civic alternative offered a positive vision for the waterfront rather than simply attempting to defeat an unpopular project. In this instance, however, the person with the power to abandon the controversial project, Donald Trump, accepted the alternative vision. Several familiar events were instrumental in effecting this outcome. The decision in an environmental suit brought by Westpride voided a zoning resolution that would have permitted Trump to build a mixed-use project at the scale he wanted. He could only negotiate down. In addition, the delays caused by the lawsuit pushed the project into the recession. New Yorkers were no longer clamoring for apartments. Trump himself was in arrears to the bank holding the mortgage on the yards, and Trump City was about to be certified by a Dinkins Planning Commission that did not look kindly on the project.[16]

On the eve of certification, Paul Goldberger gave the seal of approval to the civic alternative. "The city of New York," he wrote, "faced with the opportunity of development on an extraordinary site . . . has sat back and waited to see what would happen. It has not acted, it has reacted. . . . The [civic organizations] are doing what the city should have done years ago, which is to come forward with some vision of how this land

Figure 42. Riverside South Park, Highway-Free, Undated. (Computer Simulation by Parsons Brinkerhoff Quade & Douglas, Inc.).

should be developed to balance public and private interests. . . . The new plan for the Trump site is: practical, do–able and right for the future of New York." [17] Two months later, Trump held his first meeting with representatives from the civic organizations, and planning began for Riverside South, a project that judiciously no longer held his name. Substantially less ambitious than Trump City, the new plan would eventually be approved for 5,700 new housing units with retail, office, and commercial space and a twenty-one and one-half-acre, highway-free, waterfront park. [18] (Figure 42)

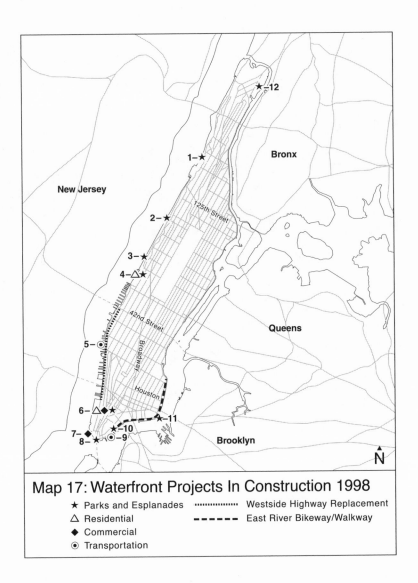

Map 17: Waterfront Projects In Construction 1998

★ Parks and Esplanades   ·················· Westside Highway Replacement
△ Residential   - - - - - East River Bikeway/Walkway
◆ Commercial
◉ Transportation

KEY

1. Riverside Park extension
   (1989–   )
2. Riverside Park renovation and
   restoration (1984–   )

3. 79th Street Marina restoration
   (1989–   )
4. Riverside South (1991–   )

258

Reaching for the Attainable, 1990–1995. The void created by the quiescent real estate market was quickly filled by proposals for waterfront parks and recreation, an unthinkable use for valuable property in times of plenty. These projects were initiated by city agencies as well as community groups. Their names— East River Docks, Chelsea Waterside Park, Riverside Walk, East River Waterside Park, and East 60th Street Pier Park— reflected their locations on the periphery of the island, areas that were no longer in demand for market use or city services.(Maps 16 and 17) Because the projects were community based and of a relatively small scale, there were a variety of donors on which to draw.[19] The timing of the 1991 federal Intermodal Surface Transportation Efficiency Act (ISTEA), with its funds for bikeways, walkways, and ferry landings, gave an added boost to the realization of these endeavors. Esplanades, sitting areas, a community center, and a small athletic field would be eked out of former Westway properties, the Banana Terminal (the last shipping pier to be cast aside on Manhattan's Lower East Side), East River Landing, and even an abandoned dumping board.

The time was finally right to revive the park at Stuyvesant Cove. On 28 February 1990, just one week after the Washington Irving hearings, the attorney for Related Companies and Deputy Mayor Fife announced the withdrawal of the Riverwalk application from ULURP proceedings. But there was a

---

5. West Side Highway replacement (1986–)
6. Battery Park City (1968–)
7. Pier A (1981–)
8. Historic Battery Park redesign (1988–)
9. Wall Street Commuter Ferry Dock (1993–)
10. Wall Street Esplanade (1992–)
11. East River Bikeway and Esplanade (1993–)
12. Sherman Creek restoration (1981–)

hitch: both Fife and Related made it clear that the developer might come back with another proposal. Community Board 6 members were undeterred, and by October, with $20,000 in hand from their Assemblyman, Steven Sanders, they had rallied the former Citizens Against Riverwalk around a plan to create an easily accessible public park at the waterfront.[20] Two years later, the city finally revoked its agreement with Related to develop Riverwalk, and signed on to the long-awaited Stuyvesant Cove park. (Figure 43)

The wasteland that was Westway also began to come alive at this time. On 1 November 1991, Roland Betts, a financier of films, chairman of the board of Sky Rink (Manhattan's only indoor ice skating facility), and the father of a figure skater, stepped onto Pier 61 on the Hudson River and determined that it would be the future home of a world class ice-skating center. The 100,000-square-foot structure, long abandoned by elegant steamships, was dark and swarming with pigeons; the roof was full of holes through which rain had drenched the concrete slab floor, creating, on this cold November day, an ersatz ice rink.

Looking upriver and down, Betts would have seen the once glorious Chelsea Piers (59 through 62, located between West 16th and 23rd streets) in a sorry state. Pier 59 was obstructed by garbage trucks and the private cars of New York City Sanitation workers. Pier 60 housed a graveyard of cars, most of them stolen, that had been impounded by the police. Pier 61 was in complete disrepair, its old, stylish windows either missing or boarded up. Pier 62 had been reduced to a concrete slab that was occasionally used for parking for a faltering excursion boat operation. Only the once pink-and-white head house had a semblance of life: soundstages had been constructed there for the weekly filming of the popular television show

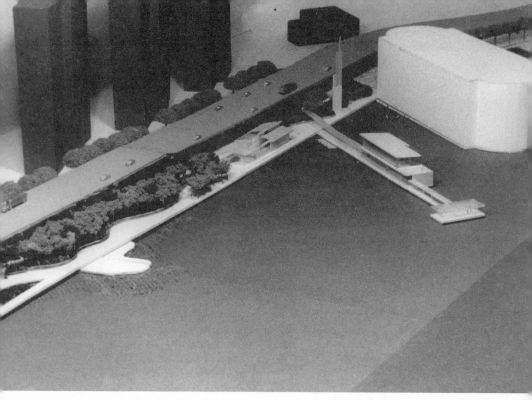

Figure 43. Model of Stuyvesant Cove, 1997, Johansson &
Walcavage, Architects. (Courtesy Ila DiPasquale).

Law and Order. A motley group of businesses used the rest of
the space for storage and parking.[21]

Betts' vision for a recreation center on the Chelsea Piers
was modest and attainable, particularly when compared to the
attempts to overhaul the West Side waterfront during the
prior decade. Unfortunately the process was little changed.
Beside the developers, four primary players would affect the
Chelsea Piers project: the state, the city, the West Side Panel,
and Community Board 4.

The state, which owned the property due to the Westway
condemnation, had an economic interest in a successful lease.
The city also had economic goals connected to this property:
first, tax revenues from upland properties would easily increase

near a thriving waterfront project; second, city officials hoped that providing more space for state-of-the-art studios would reinvigorate New York film production, which was ailing after a recent strike.[22] In addition to these economic factors, the promotion of a successful development had personal appeal for the city's chief executive: Mayor Dinkins had been a supporter of the recreational renewal of this West Side waterfront since his borough president days.[23]

Despite these positive auguries at the state and local level, an auction held just a month after Betts' initial visit to the pier failed to attract any serious offers, due largely to the short duration of the permit term: a mere five years, with a ten-day termination clause. This short-term permit was not conducive to financing for such a large-scale project. Furthermore, the state inserted a clause in the permit language that held the permit holder responsible for all repairs, the extent of which were unknown.[24]

Meanwhile the official West Side Waterfront Panel mounted a challenge against the proposed development. This was, after all, its role. The panel's widely supported Hudson River Waterfront Park plan had become the policy document for the West Side waterfront. This plan asserted that public service uses, such as sanitation garages and tow pounds, should be removed from the waterfront as soon as financing could be arranged and other appropriate locations could be found. Betts' proposal was at odds with some elements of this plan, such as the mandate to maintain sight-lines to the Hudson on all east-west streets. Betts wanted to keep the Chelsea Piers head house intact, even though it would block nearly a quarter mile of views and physical access to the waterfront.[25]

There were also turf concerns. The city and state were dawdling over the creation of a joint entity to replace the panel and to implement the park plan. Meanwhile, without a control mechanism to protect the park, how could the state lease property that was intended to be a part of the plan?

The residents of Community Board 4, in whose district the Chelsea Piers are located, would clearly benefit if Betts' ultimate vision for the piers were to be realized. There was still a dearth of recreational facilities, and the waterfront remained walled off. Initially, however, these residents too had reservations. They had long wanted to expand a leftover traffic triangle at West 23rd Street and convert it to open space. They envisioned Chelsea Waterside Park as part of the Hudson River park extending out onto Pier 62. An auction that put the piers in the wrong hands might jeopardize these plans.[26]

Betts sought a two-rink facility that would operate on Pier 61 twenty-four hours a day, year round. It would serve the general public, Olympic caliber figure skaters, and hockey players for a reasonable fee. It would also offer training opportunities for the disadvantaged. The pier would be restored to its former glory, with windows on three sides looking out on the George Washington Bridge, the Statue of Liberty, and the "best sunset views in Manhattan."[27] Betts, new to development in New York, and particularly to development on the waterfront, had no idea what was in store for him.

Shortly after his November resolution, Betts hired his college roommate, James G. Rogers III of Butler Rogers Baskett Architects, to give form to his concept. Within a month the architects had drawn up a plan. Betts and his partners, David Tewksbury (president of Sky Rink) and Tom Bernstein (Betts' partner in his film venture, Silver Screen Management), began to court the people in power. The partners needed support for a ten-year, renewable lease. They spent December 1991 making presentations to elected officials, members of the West Side Waterfront Panel, and community and citywide organizations.

Nearly everyone they approached was supportive, except Tom Fox, a panel member who would soon become the president of the Hudson River Park Conservancy, the panel's successor entity. Fox saw an opportunity to derive income from an abandoned waterfront structure at no cost to the city or state, and to jump-start the stalled Hudson River park. Yet Fox wanted even more. He hoped to rid Piers 59 and 60 of the sanitation and tow pound facilities. Because neither relocation funds nor alternative sites were available, Fox needed another way to clear those piers. He arranged for an engineering survey, which found the superstructures to be unstable. The state then gave the city thirty days to vacate the condemned facilities—no strings attached.[28]

Community Board 4 pressed for an even broader vision for the piers. In an early January resolution, the board endorsed the Sky Rink proposal but called for more public access, as well as a sports center with tennis courts, bowling alleys, indoor gyms for families, day care centers, golf programs, rock climbing, batting cages, and more. They also recommended a marina for small boat rentals, a sailing and fishing school, and acceptable satellite retail operations.[29]

Although most city officials believed that the Sky Rink proposal was "pie in the sky," by mid-January the state announced the formation of a Task Force to prepare bid documents for a March auction of the entire Chelsea Piers complex with a five-year, one-time renewable lease term.[30] Fox had prevailed in his desire to link the development of all four piers. Now, instead of dealing with a single pier, Betts and his partners were faced with a huge project (the equivalent of four ninety-story skyscrapers lying on their sides, covering four piers and the head house). And the lease term was still too short to secure financing. Nonetheless, they decided to go for it.[31]

Their final proposal for the Chelsea Piers, submitted to the state in May 1992, satisfied the interests of all parties and came well within bid expectations. Sky Rink would move to a twin-rink facility on Pier 61. Nine additional sound stages would be constructed as an expanded home for Betts and Bernstein's Silver Screen enterprise. A full service marina would be located at Pier 62, with additional marina space on wharfage throughout the property. Pier Park, with seating, wind-tolerant "beach" plants, and fabulous views, would be expanded to a public esplanade along the apron of all four piers. Possible future projects would include many of the recommendations put forward by the community and potential tenants. In his cover letter, Betts stressed his commitment to the West Side Waterfront Panel's vision of a "waterfront . . . [that was] open, accessible and dedicated primarily to public recreation and maritime activity." [32]

The following month, the state awarded a renewable ten-year lease to Betts' Chelsea Piers Management. Betts immediately assembled a design and development team. He also began the tortuous process of acquiring the necessary permits.

Although the city was in the process of rationalizing the permit system, not much had changed twenty years after the opening of the River Café. In the 1880s, Chief-Engineer Green had struggled with the U.S. Army Corps of Engineers, the state legislature, and the City Board of Estimate to build the Chelsea-Gansevoort piers. Betts' task seemed formidable in contrast. The Chelsea Piers property was subject to city zoning and code regulations. Thus, Betts needed waivers from the Board of Standards and Appeals (BSA) for such things as outside staircases used as fire escapes. These structures had been allowed when the piers were used for passenger access to ocean liners, but, like River Café, violated fire codes for land based buildings, which ironically now also applied to the piers.

Because the site was still owned entirely by the state, it was also subject to regulations issued by the departments of Conservation and Transportation. Two of the piers were also on the state's Historic Structures list, which meant that any changes required the approval of the Office of Parks, Recreation and Historic Preservation.

All four of the piers stand in a federal navigable waterway, making any alterations subject to review by the U.S. Army Corps of Engineers. In addition, because the project was deemed to have a potentially negative effect on the environment, it fell under the National Environmental Policy Act and thus required an EIS.

Not only did Betts have to wend his way through a labyrinth of agencies, but the various approval processes also required public hearings and input from the Community Board, elected officials, and the general public. In fact, the imminent Chelsea Piers project had itself been a catalyst for establishing yet another bureaucratic entity to which Betts needed to argue the merits of his plan. Just five days before the state auction (after four months of inaction), the city and state had signed a Memorandum of Understanding creating a management entity for a West Side waterfront park: the Hudson River Park Conservancy.

Perhaps inexperience with waterfront projects and naïveté were Betts' saviors. He plowed ahead, relying on a well-organized plan, a rational manner, and excellent personal contacts. He simply could not afford to wait. In October 1992, just five months after the state award, the final EIS process started. Shortly thereafter he signed an interim lease with the state, and began work with the BSA. For fourteen months, Betts held the city agencies to a timetable of monthly hearings. When agency officials came to these hearings unprepared, Betts reached out to Dinkins, who assigned his Deputy

Mayor, Norman Steisel, to act as an expediter. In December 1993, little more than a year after its start, the final EIS was certified; four months later the BSA waivers were also granted. In September 1994, some twenty-seven months after having been awarded the project, Betts began construction on the Chelsea Piers.[33]

Financing of the project, another major hurdle, proceeded simultaneously. Other than Sky Rink, there was simply no precedent for such a grand sports complex in New York City. Potential investors were difficult to find, even with the longer lease term Betts won from the state. In their auction bid, the Chelsea partners had guaranteed they would spend $50 million to renovate three of the piers, with an option to add a fourth (Pier 60). To satisfy this guarantee, Betts, Bernstein, and Tewksbury decided not to seek public assistance, which would ensure additional applications and procedures, and inevitable delays. Instead, they created a Limited Partnership and contributed $17 million of their own funds. Morgan Stanley also sold junk bonds for the project, bringing the initial capitalization to $62 million.

When unanticipated construction costs (including the eventual acquisition of Pier 60) pushed the project considerably over budget, Betts' personal contacts once again became a valuable asset. Calls to influential friends and the addition of tenant-financed capital improvements brought the entire investment in the piers, raised over seventeen months, to $110 million. Not a dollar of that money was from public sources.

The Chelsea Piers opened on 25 October 1995, a somewhat remarkable four years after Betts' initial visit to Pier 61. Decades after the *Wateredge Development Study*, Westway, and Trump's first proposal for the Penn railroad yards, no roadway, park, or housing construction had begun; at the Chelsea Piers, however, recreation had finally made an inroad on the waterfront. The complex may not be the passive, luxuriant, green space that some wanted, but the public can now walk, unimpeded by traffic, along the edge of the Hudson River in a setting that reflects its maritime heritage. At several points, roadways through the head house open sight-lines from the inner city to the water, while roll-up doors on the piers offer views to dozens of excursion and recreational vessels. (Figure 44)

There were two factors that contributed to the success of the Chelsea project: it was centered on recreation at a time when the Hudson was clean and its potential recreational benefits widely recognized, and, perhaps more important, all the interested parties genuinely wanted the project to succeed. In the end, sheer force of will, a readiness to be energized rather than discouraged by frustration, political savvy, and the ability to secure private financing finally broke down a portion of the walls that had closed off Manhattan's waterfront for so long.

All At Last Returns to the Sea . . . the Beginning and the End, 1995–1998.[34] As the millennium approaches, there are hopeful signs that the Manhattan waterfront will at long last emerge from its degenerated state. Battery Park City is building again: the Holocaust Museum opened on the south end in 1997, and additional apartment houses, a school, and a hotel are underway, following the new guidelines for the northern sector. After years of waiting for permits and financing, construction has finally begun on a restaurant and visitors center on Pier A, the former Battery home of the Department of Docks. There are plans for two new commuter ferry terminals to serve the reawakened interest in this historic mode of waterborne travel. And, in the same month that New Orleans announced the demise of its riverboat gambling casino, New York City—

Figure 44. Chelsea Piers, 1996. (Aerial Photography by Beck's Studio).

fresh from victory in the courts—announced new hope for the Passenger Terminal piers. Gambling boats would soon depart from the piers to waters just three miles beyond the city limits.[35]

Additional sections of Manhattan's waterfront continue to open up to the public. Along several blocks of Route 9A to the west of Greenwich Village, a tree-lined boulevard (a reminder of 1930s proposals for the East River Drive) with a walkway and bikeway on the border closest to the Hudson is complete. Financing came from Westway trade-in funds. By the end of the decade, this boulevard will extend northward beyond West 23rd Street and other esplanades and bikeways—beneficiaries of ISTEA—will fill in missing links along the East River from Battery Park through Stuyvesant Cove. To the west of Route 9A, the first section of what is officially called Hudson River Park is under construction. New York's Governor George E. Pataki even signed legislation acceptable to the city and state defining its boundaries and governance.[36] Further up the Hudson, the Parks Department is creating waterfront walkways and kayak launches where it once allowed only ballfields and tennis courts. And in northern Manhattan along the Harlem River, the actress Bette Midler has organized a group of volunteers to restore debris-laden Sherman Creek (in the 1980s slated for a sewer plant) to its natural state. (Map 17)

But the hope engendered by these actions is tinged with despair. At Stuyvesant Cove, State DOT and ISTEA funds have been allocated to build a park, and there are designs that include a bikeway/walkway. But underwriting to plant trees or to build a nature center intended to generate maintenance funds remains elusive, nor has it been decided who will own and maintain the park.

There continues to be concern over the ultimate configuration of Hudson River Park. The Conservancy, utilizing funds from the rental of some still-standing Westway properties, created temporary esplanades and shored up several piers. Its objective was to draw New Yorkers to the Hudson in order to help them understand the potential of its constantly degenerating shore. While the new law recognizes their eventual removal, several piers slated to become part of the park are still being used to store garbage trucks, police barricades, and impounded cars. Without an alternative site and adequate funds to move these obstacles, the riverfront will remain blocked for the foreseeable future.

Further uptown on the West Side, construction began in the winter of 1997 on two of the six anticipated Riverside South apartment towers; concurrently children could be seen playing soccer on temporary lawns along the shore. Work on a recreation pier that until the 1990s had been reserved for municipal uses, and on the first nine acres of the park (from West 73rd to West 68th Street), began in 1998. The Miller Highway casts a formidable shadow on the park, however, and although the EIS for its removal is nearly complete, there are no construction funds in sight.[37]

There is also hope for the southern tip of the island. In the early 1990s, a fire set by homeless squatters caused significant damage to the South Ferry site, which the developers had returned to the city. Officials then revised their approach to this site and pursued a more modest plan to replace only the ferry terminal. An international competition was held; Robert Venturi's winning design was for a low-rise structure with "a clock scaled to lower Manhattan as a whole and to its harbor—a clock bigger than those of numerous 19th century railroad terminals in America or Big Ben in Westminster—a clock 120 feet in diameter appropriate in scale . . . to our collective-hype-sensibility today—the largest clock in the world."[38] What the *AIA Guide to New York* once described as "the world's most banal portal to joy (a public men's room en route to Mecca)" was to become a "celebratory place that uplifts the spirits upon arrival or departure . . . a symbolic portal to Manhattan [that] will assume a place among New York's landmark destinations."[39]

Once again, however, what appeared to be an exciting step forward for the waterfront has been slowed by controversy.

The clock design was not universally popular, and political pressures from Staten Islanders (who had helped to elect the new mayor, Rudolph Giuliani) caused the design to be altered so radically that Venturi quit the project. The final design for the terminal is quite pedestrian. Nonetheless, a new ferry terminal is still in the works.[40]

Although much of the negativism engendered by Westway has been countervailed and the opposition has weakened, the West Side waterfront projects currently on the drawing board are by no means sure things. At the EIS hearings on Riverside South, one commentator remarked that the project offered the "possibility of breaking the cycle of recrimination, suspicion and hostility that has so often surrounded major development projects in New York."[41] That possibility has not turned into reality. The Coalition for a Livable West Side, a group of no more than a half dozen tenacious fighters, continues to join every lawsuit, to testify at every opportunity, and to reach out to the press, which frequently prints articles favorable to its cause.

Congressman Jerrold L. Nadler, aided by other Upper West Side elected officials, used his power in Washington to further the protestors' cause.[42] In actions reminiscent of those the New Jersey legislators used to block Westway, Nadler managed to sidetrack and reduce the ISTEA appropriation needed to conduct the environmental studies required for the highway relocation. His actions were supported by a city administration that had a different use in mind for the funds: highway repairs in boroughs where votes were needed, and a new Pennsylvania Station.

In the former Westway corridor, the city and the state have both provided their share of stumbling blocks to the creation of Hudson River Park. Not until the 1996 Environmental Bond Act was passed (with heavy support from environmentalists and parks advocates) did Governor Pataki finally commit the state's share of the $200 million promised by former Governor Cuomo and former Mayor Dinkins to build the park. During the Dinkins administration, the city had been on board with both funds and spiritual backing. Under Giuliani, it fought to retain its properties north of 42nd Street for non-park uses—in order to take advantage of the upswing in the real estate cycle that began in the mid-1990s. The new park law conceded four piers to the city, removing them from the park.

On a more positive note, former environmental opponents were organized by Westway litigator Al Butzel into a coalition to "create a continuous world-class park along the Hudson in New York City." This move leaves opposition to Hudson River Park in the hands of a vocal, self-interested group that calls itself The Federation to Preserve the Greenwich Village Waterfront and Great Port. These activists want the waterfront to remain exactly as it is. In addition, the respected New York Public Interest Research Group, the New York City chapter of the Sierra Club, and Marci Benstock—who has become a virtual one-person Clean Air Campaign—still see both the Conservancy and its replacement Trust as a reincarnation of Westway, out to create a development park.[43] In a 1997 statement commenting on the possibility of creating beaches along the Hudson, Benstock showed just how radical she had become. Claiming that any "intrusion" into the water to allow swimming would threaten the habitat of more than 100 species of fish she added, "things that don't have to be in the water don't make sense for any aquatic habitat."[44]

Although Manhattan's waterfront has come a long way from the shores of Nieu-Nederlandt, the issues in the twenty-first century will not differ significantly from those that have evolved along with the layers of land, wood, and concrete.

273

Vocal groups continue to battle for and against any proposed mode of waterfront rejuvenation. The question now is, how can an opposition that has become so firmly entrenched be helped to save face and back off? The time it takes to accomplish anything on the waterfront may speed up with new zoning regulations and streamlined agency reviews. But these reforms have yet to be tested, and the danger always remains that a new administration will refuse to honor projects unless they are fully funded or already under construction. More than 300 new acres could potentially be added to the city park system, but issues of ownership, maintenance responsibility, and (above all) funding for upkeep remain unsolved. A constituency has to be nurtured to move park blueprints off the drawing boards and into open space construction.

The plans put forth in response to Engineer-in-Chief George McClellan's invitation in 1870 showed rows of decorative warehouses. In their time, the Chelsea-Gansevoort piers were the acme of pier design. Today, a constant shortfall of funds and the perceived need to meet a common denominator or to conform to the politics of taste often prevent the achievement of high quality design. In order to reach consensus, the Whitehall ferry terminal planners succumbed to a plain vanilla coating.

At Riverside South, the dilemma is positively Machiavellian. Here is the opportunity to build a waterfront park that is a model for the twenty-first century. Still, the highway looms overhead. (Figure 45) Many activists actually believe that the park needs to be attractive enough that people will be drawn to visit the site (after which they will presumably understand the need to move the highway), but that at this stage there is no need to create a truly beautiful park that might mobilize a constituency that would fight to make the park—highway and all—permanent.

Figure 45. Riverside South Park with Miller Highway in Place, Undated. (Computer Simulation by Parsons Brinkerhoff Quade & Douglas, Inc.).

Landfill, so much a part of Manhattan's heritage, will likely return in new guises. The need to provide walkways in parks with an extremely narrow edge, and to connect esplanades where highways and apartments still intrude on the river, provides a compelling argument to expand the island's borders in a controlled manner. There may also be an opportunity to create new recreation land while at the same time providing a means to dispose of the tons of material dredged from the harbor annually to accommodate shipping.

Much of Manhattan's waterfront, with its damaged and abandoned layers, is entering the twenty-first century in a tawdry state: chain-link fences, buckled piers with huge chunks missing, a landmark former ferry landing in need of more than $30 million in repairs, tennis bubbles blocking views of the Brooklyn Bridge, parking lots, sand piles, temporary roadways, and highways often on the verge of collapse. Despite its sorry condition, however, people are increasingly bound for the water's edge, to ride new ferries to work, to sit and gaze, to bike and rollerblade, to sail and kayak, to fish, and even to swim. Historically New York has a poor record of keeping its shores open for its citizens, and the threat of privatization becomes stronger with each rebound of the economy. After years of false starts and piecemeal rewards, we must search assiduously for the means to give Manhattan's waterfront back to its people.

NOTES

1. City of New York Parks & Recreation, *Waterfront Management Plan*, 1990.
2. The earthquake put an end to periodic voter referendums over completing the unfinished highway. Material on waterfront developments nationally and internationally during this period may be found in *Waterfront World;* Ann Breen and Dick Rigby, *Waterfronts, Cities Reclaim Their Edge,* (Washington, D.C.: MacGraw-Hill, Inc., 1994); and Breen and Rigby, *The New Waterfront, A Worldwide Success Story,* (Washington, D.C.: MacGraw-Hill, Inc., 1996).
3. In 1988, President Bush visited Boston and declared the harbor the nation's dirtiest. Work on Deer Island began the following year. *Waterfront World* (Summer/Fall 1996): 5.
4. The $55 billion value of cargo handled by the Port of Los Angeles in 1990 superseded the $49 billion activity in the Port of New York.
5. In the late 1980s, in order to deal with overcrowded local jails, the city

purchased three floating dormitories that had been mothballed after the Persian Gulf War. These barges were stationed at Pier 40 in the Hudson, and adjacent to the Banana Terminal on the Lower East Side. Needless to say, unlike the reaction to the earlier floating pools, community opposition was intense.

6. Peter Newell, "Rivercrawl," *East Side Express,* 10 March 1983, 1, 6; and Frank Sommerfield "17 Acres, River View," *7 Days,* 15 March 1989, 41.

7. *578 Miles of Opportunity, New York City's Waterfront,* Recommendations from 1984 Waterfront Conference, 1985, 3. The Regional Plan Association was a major participant in this conference.

8. Fife was Chair of the Parks Council's Waterfront Committee from 1981 to 1985. The author succeeded her as chair of this committee.

9. City of Boston, Boston Redevelopment Authority (BRA), *A Framework for Planning Discussion,* October 1984. In October 1990, BRA published *The City of Boston Municipal Harbor Plan* that expands on the earlier framework and includes zoning text amendments. Royal Commission on the Future of the Toronto Waterfront, 1992.

10. The city's waterfront plan uses this terminology. The total circumferential mileage depends on whether the measurement is taken with or without piers. From 1990 to 1994, the author was vice president for Waterfront Planning and Development at the *Economic Development Corporation (EDC).* She took part in both the City Planning and Borough President's planning activities. During this period, she also participated either as a city official or as a Parks Council vice president, in waterfront park projects.

11. *New York City Comprehensive Waterfront Plan, Reclaiming the Edge,* 1992; *Amendment to the Zoning Resolution of the City of New York,* August 1993; and *Comprehensive Manhattan Waterfront Plan,* 1995, approved by Resolution No. 2291 of the City Council as a 197A Plan on 16 April 1997.

12. The rules are more stringent for Manhattan. The Borough President's Plan completely disallows landfilling and platform development on pilings. Section 382-a of the 1990 New York State omnibus tax bill, prohibits new landfill, platforms, pilings, piers, or structures in the Hudson between Battery Park City and 35th Street.

13. Letter from Michael J. Del Giudice, chairman, to Governor Mario M. Cuomo and Mayor David N. Dinkins, 1 November 1990; and West Side Waterfront Panel, *A Vision for the Hudson River Waterfront Park,* 1 November 1990, 9, 17, Author's collection, File, "West Side Task Force."

14. Gutman, interview.

15. Eventually the Natural Resources Defense Council, Westpride, and the Riverside Park Fund joined the consortium.

16. Robinson, interview.
17. Goldberger, *NYT,* 1 July 1990.
18. The Riverside South Planning Corporation, consisting of the civic groups (nicknamed the "civics") and the Trump Organization, was formed in May 1991. In return for the civics' support through ULURP, Trump agreed to set up a Board of Directors that would reach decisions by consensus and engage in a collaborative design process. Richard Kahan, interview by author, 10 October 1997.
19. These schemes were put forth by local and citywide park groups as well as the Department of Parks & Recreation and the Economic Development Corporation, successor to the Public Development Corporation. Funding sources were as follows: East River Docks (renamed Wall Street Pier and Esplanade), EDC and ISTEA; Chelsea Waterside Park, New York State DOT mitigation funds; Riverside Walk, Manhattan borough president and ISTEA; East River Waterside Park, developer open space mitigation required by the Department of Planning; East 60th Street Pier Park, developer mitigation and private donations.
20. *Programming Requirements for Waterfront Open Space Study,* "Stuyvesant Cove." An ardent supporter of Citizens Against Riverwalk was made chair of the Park Committee. Crotty, interview.
21. Roland Betts, interview by author, 23 July 1997; author's own recollections.
22. Memorandum from Lance Ruiz Carlisle to Carl Weisbrod, 18 January 1991, Author's collection, File, "Chelsea Piers."
23. Dinkins was a member of the West Side Task Force, which recommended establishing the park.
24. New York State DOT, Undated, "Proposal and Occupancy Requirements," Author's collection, File, "Chelsea Piers;" and Memorandum from Kim Ile to Lance Ruiz Carlisle, 14 January 1990, Author's collection, File, "Chelsea Piers."
25. West Side Waterfront Panel, 1990; and Betts, interview.
26. Thomas Balsley, Landscape Architect, interview by author, 27 October 1997.
27. Betts, interview.
28. Betts, interview; and Tom Fox, interview by author, 14 August 1997.
29. 8 January 1992, Community Board 4, File, "Chelsea Piers."
30. Memorandum from Lance Ruiz Carlisle to Carl Weisbrod, 18 January 1991; and Memorandum from Ann Buttenwieser to Files, 17 January 1992, Author's collection, File, "Chelsea Piers." The author was a member of this Task Force representing EDC.
31. New York State DOT, *Bid Package, for Public Auction to be held Friday,*

May 2, 1997; Chelsea Piers Sports and Entertainment Brochure, 1997; and Betts, interview.

32. Chelsea Piers, L.P., *Chelsea Piers Proposal, Recreation and Entertainment Complex,* 22 May 1992.

33. Descriptions of the approval process and financing are from Betts interview.

34. Rachel Carson, *The Sea Around Us* (New York: Oxford University Press, 1951) 14, ending.

35. Ruling by U.S. District Court Judge Allyne Ross that a Brooklyn U.S. Attorney's opinion that gambling boats must go offshore twelve miles was too broad an interpretation of federal law.

36. State of New York, 9833-B, in assembly, 10 March 1998. Under this legislation the conservancy will be replaced by the Hudson River Park Trust, an independent public benefit corporation made up of thirteen members appointed by the governor, the mayor, and the Manhattan borough president. The trust has the broad powers to plan, design, develop, construct, operate, and maintain the park. The park's name has been officially shortened from Hudson River Waterfront Park to Hudson River Park.

37. At about the same time that the civic organizations formulated their plan to move the highway, state engineers checking its structural integrity found it unsafe. The civics pressed to use the $70 million allotted for repair for removal instead. The timing was off by six months. Riverside South had not been approved by ULURP and the state insisted that it could not wait. The civics sued on the grounds that an EIS was required to consider alternatives to repair. But the suit was aborted in return for ULURP approval of the residential project. Gutman, interview; and U.S. District Court, Southern District, 90CIV7904KTD, filed 1990 December.

38. Venturi, Scott Brown and Associates and Anderson/Schwartz Architects, *A Design for the Whitehall Ferry Terminal,* October 1992, EDC, File, "Whitehall."

39. AIA Guide to New York, Collier Books, 1978, 7; and *Whitehall Ferry Terminal Competition,* Charge to the architects, August 1992, EDC, File, "Whitehall."

40. No sooner was the clock eliminated from the design than its chief critic, Staten Island Borough President Guy Molinari, announced plans for an otherworldly terminal designed by architect Peter Eisenman on the Staten Island side of the harbor. Clifford J. Levy, "Not Just a New Ferry Terminal, but a Fanciful One," *NYT,* February 1997.

41. *Final Environmental Impact Statement Riverside South,* Response to comments, October 1992.

42. Others actively engaged in the battle were state assemblyman Scott Stringer, state senator Franz Leichter, and councilwoman Ronnie Eldridge.
43. Butzel, interview.
44. Douglas Martin, "Swimming in the Hudson? Maybe," *NYT,* 11 August 1997, B2.

Select Bibliography

Documents

Allee King Rosen & Flemming. *Riverside South. Final Environmental Impact Statement,* 1992.

Borough President of Manhattan. *Annual Reports.* New York, 1902–42.

Common Council. *Minutes,* 1800–31.

Craig, Charles L. *Riverside Park Improvement,* 30 January 1924.

Henry Hudson Parkway Authority. *Opening of the Henry Hudson Parkway and Progress on the West Side Improvement.* New York, 12 December 1936.

Minutes of the Common Council of the City of New York 1760–90. 8 vols. New York: Dodd Mead, 1905.

Moses, Robert. Memorandum to the Mayor on Park Department Revised Plan for West Side Improvement in Riverside Park. New York, 10 June 1935.

New York City. Board of Aldermen. *Documents of the Board of Aldermen of the City of New York.* 35 vols. New York, 1835–1870.

———. *Proceedings.* Vols. 8–216. New York, 1834–94.

———. *Report of the Special Committee on Parks, Relative to Laying Out a New Park in the Upper Part of the City.* Doc. No. 83, 2 January 1852.

———. *Report of the Committee on Wharves Relative to the Erection of a Great Pier in the North River.* New York: William B. Townsend, 1836.

New York City. Board of Assistant Aldermen. *Documents.* New York, 1834–50.

New York City. Board of Estimate and Apportionment. *Minutes,* 1910–40.

New York City Inspector. *Annual Reports.* New York, 1857–73.

New York City. Department of Docks. *Annual Reports.* New York, 1870–1935.

———. *Minutes,* 1870–1935.

———. Plan Collection, 1870 to present.

New York City. Department of Finance. Bureau of Real Estate. *Record of City Owned Property,* 1700–1900.

New York City. Economic Development Corporation and Department of Transportation. *Whitehall Ferry Terminal Design Competition,* 1992.

———. *Whitehall Ferry Terminal Project,* 1992.

New York City. Improvement Commission. *Report to the Honorable George B. McClellan Mayor of the City of New York and to the Honorable Board of Aldermen of the City of New York.* January 1907. New York: Kalkhoff, 1907. (Preliminary Report 14 December 1904.)

New York City. Municipal Archives. Annual Record of Assessed Valuation of Real Estate, 1880–1940.

———. Common Council Papers, 1830–50.

———. Department of Docks. Letters, 1870–90.

———. Department of Docks. Glass Negative Collection, 1870–1935.

———. Mayors Papers, 1910–40.

———. Photographs. Manhattan Borough President and Department of Public Works.

———. Works Progress Administration. Federal Writers' Project. NYC Unit. Photographs.

New York City. Office of the Mayor. *Descriptive Features East River Drive: Grand Street to East 60 Street,* 1936.

New York City. Parks Department. *Annual Reports,* 1871–1920.

New York County. Register's Office. Re-Indexed Conveyances of the Blocks and Lots of the City of New York . . . Prior to 1917.

New York New Jersey Port and Harbor Development Commission. *Joint Report: With Comprehensive Plan and Recommendations.* Albany: J. B. Lyon, 1920.

New York State. Department of Transportation and Office of Parks, Recreation and Historic Preservation. *Draft Report on Conceptual Design Alternatives for Westway State Park,* 1983.

New York State. Department of Transportation and Office of Parks, Recreation and Historic Preservation. West Side Highway Project. *Conceptual Design Alternatives for Westway State Park,* 1984.

New York State. *Report of the Commissioners of the Land Office Relative to New York Harbor Encroachments.* Sen. Doc. 10, 9 January 1862.

New York State. *Report of the New York Harbor Commission of 1856 and 1857.* New York: C. S. Westecott, 1864.

Parks Council. *Agenda for Action. 578 Miles of Opportunity, New York City's Waterfront.* Recommendations from 1984 Waterfront Conference, 1985.

Public Meetings of the Department of Docks to Hear Persons Interested in Improving the River Front, June and July 1870. New York: The New York Printing Company, 1870.

Report of the West Side Improvement Architects' Committee, 29 April 1929.

Report of the West Side Improvement Engineering Committee, 13 May 1927.

Tompkins, Calvin T. *Department of Docks.* New York: M. Brown Printing, 1910.

United States Congress. House Committee on Government Operations. *The Westway Project: A Study of Failure in Federal/State Relations.* Washington: U.S. Government Printing Office, 1984.

U.S. Department of Transportation. Federal Highway Administration. New York State Department of Transportation. *Final Environmental Impact Statement/Design Report/Section 4(f): Evaluation for Route 9A Reconstruction Project Battery Place to 59th Street, New York County, New York,* 1994.

Select Bibliography

Private Documents

Columbia University. Manuscript Collection. Dorothy Rosenman Papers.

———. Lillian Wald Papers.

Committee of Seventy. *Report of Sub-Committee on Improvements of the Waterfront.* New York, 1895.

The New-York Historical Society. Manuscript Collection. Astor Papers.

———. Women's League for the Protection of Riverside Park Papers. 1916–1938.

New York Public Library. Manuscript Collection. Stanley Isaacs Papers.

Title Guarantee and Trust Company. Title Searches.

University of Minnesota. Social Work History Archives. Henry Street Settlement Collection.

West Side Association. *Proceedings,* 1871–74.

Contemporary Sources

Bonner, William Thompson. *New York The World's Metropolis 1623–4—1923–4.* New York: R. L. Polk, 1924.

Bradley's Reminiscences of New York Harbor. New York: David L. Bradley, 1898.

Bradley's Water Front Directory. New York: David L. Bradley, 1881.

Citizens Association of New York. *Report Upon the Sanitary Condition of the City.* New York: D. Appleton, 1866.

DeVoe, Thomas F. *The Market Book Containing a Historical Account of the Public Markets of the City of New York.* New York: Printed for the author, 1862.

Gothams Great Rotten Row: Peter B. Sweeny's Project for a Splendid Public Pleasure Ground for Lovers of the Horse and the Horse Himself—A Grand Terrace on the West Side . . . New York: Municipal Improvement Association, 1890.

Haswell, Charles H. *Reminiscences of an Octogenarian.* New York: Harper and Brothers, 1896.

Hawkins, Rush C. *Corlears Hook in 1820.* New York: J. W. Bouton, 1904.

Henry Street Studies. *A Dutchman's Farm: 301 Years at Corlears Hook.* New York: Henry Street Studies, 1939.

Hoffman, Murray. *Treatise Upon the Estate and Rights of the Corporation of the City of New York as Proprietor.* New York: McSpedon and Baker, 1853.

Lewis, Nelson. *City Planning Pamphlets,* n.p. 1915.

New York Pier and Warehouse Co. *Piers and Wharves of New York.* New York: Evening Post Steam Presses, 1869.

Parkhurst, Charles. *Our Fight With Tammany.* New York: Scribners, 1895.

Poole, Ernest. *The Harbor.* New York: Macmillan, 1915.

Regional Plan of New York and Its Environs, The. 8 vols. New York: Regional Plan of New York and Its Environs, 1929; reprint ed., New York: Arno Press, 1974.

Riis, Jacob A. *How The Other Half Lives: Studies Among the Tenements of New York.* 1890. New York: Dover, 1971; reprint ed.

Rikeman, A. A. *The Evolution of Stuyvesant Village.* New York: Curtis G. Peck, 1899.

Smith, R. A. C. *The West Side Improvement: Its Relation to all of the Commerce of the Port of New York,* 7 June 1917.

Sweeny, Harry Jr., ed. *Opening of the West Side Improvement.* New York: Moore Press, 1937.

Trows. *City Directory,* n.p. 1870–95.

Valentine, David T., ed. *Manual of the Corporation of the City of New York.* 30 vols. New York: David T. Valentine et al., 1841–1871.

Viele, Egbert. *The Topography and Hydrology of New York.* New York: Robert Craighead, 1865.

Waring, George E. Jr. *Street Cleaning and the Disposal of a City's Waste: Methods and Results and the Effect Upon Public Health, Public Morals and Municipal Property.* New York: Doubleday and McClure, 1898.

Zeisloft, E. Idell. *The New Metropolis.* New York: D. Appleton, 1899.

Maps

Bridges, William. *Map of the City and Island of Manhattan with Explanatory Remarks and References.* New York: T. Swords, 1811.

Select Bibliography

Bromley, G. W. *Atlas,* 1890–1940.

Burgis View of the "South Prospect of the Flourishing City," 1717.

Gerard, J. W. Jr. *A Treatise to the Title of the Corporation and Others to the Streets, Wharves, Piers, Parks, Ferries and Other Lands and Franchises in the City of New York.* New York: Baker Voorhis, 1873.

Janvier, Thomas A. *In Old New York.* New York: Harper and Brothers, 1894.

Perris, William. *City Insurance Maps,* 1870–90.

"Plan of the City of New York Showing Fire Districts, etc." New York: Narine & Co., 1839.

Robinson, E. *Atlas of the City of New York,* 1880–1910.

Viele, Egbert L. "Sanitary and Topographical Map of the City and Island of New York," 1865.

War Department Corps of Engineers, U.S. Army and United States Shipping Board. *The Port of New York.* Washington D.C.: U.S. Government Printing Office, 1926.

Willis, Bailey and Dodge, R. E. "Physiographic Features of the District." New York Geological Survey, 1901.

Newspapers and Journals
American City Magazine. 1910–40.
Architectural Record. 1963–86.
East Side Chamber News. 1929–40.
Harper's Weekly. 1870–90.
New York Herald. 1870–1940.
New York Times. 1870–1998.
Municipal Affairs. 1897–1900.
Municipal Engineers Journal. 1900–40.
Progressive Architecture. 1966–86.
Public Improvements. 1898–1900.
Real Estate Record and Builders' Guide. 1870–1940.
Regional Plan Association. *Bulletin.* 1931–40.
Scientific American. 1880–1910.
Survey, The. 1931.

Secondary Works

Albion, Robert Greenhalgh. *The Rise of New York Port 1815–1860.* New York: Scribners, 1939.

Black, George Ashton. "The History of Municipal Ownership of Land on Manhattan Island," *Columbia University Studies in History, Economics and Public Law,* vol. 1, 1897.

Breen, Ann and Rigby, Dick. *The New Waterfront, A Worldwide Success Story.* Washington, D.C.: McGraw-Hill, Inc., 1996.

———. *Waterfronts, Cities Reclaim Their Edge.* Washington, D.C.: McGraw-Hill, Inc., 1994.

Bunker, John G. *Harbor and Haven.* Woodland Hills, Calif.: Windsor, 1979.

Caro, Robert A. *The Power Broker: Robert Moses and The Fall of New York.* New York: Knopf, 1974.

Condit, Carl W. *Chicago 1920–29.* Chicago: University of Chicago Press, 1973.

———. *Chicago 1930–70.* Chicago: University of Chicago Press, 1973.

———. *The Port of New York: A History of the Rail and Terminal System from the Beginnings to Pennsylvania Station.* Chicago: University of Chicago Press, 1980.

———. *The Port of New York: A History of the Rail and Terminal System from Grand Central Electrification to the Present.* Chicago: University of Chicago Press, 1981.

Cranz, Galen. *The Politics of Park Design: A History of Urban Parks in America.* Cambridge: MIT Press, 1982.

Fein, Albert, ed. *Landscape into Cityscape: Frederick Law Olmsted's Plans for a Greater New York City.* Ithaca: Cornell University Press, 1967.

Gilder, Rodman. *The Battery.* Boston: Houghton Mifflin, 1936.

Grafton, John. *New York in the 19th Century.* New York: Dover, 1977.

Griffin, John I. *The Port of New York.* New York: Arco, 1959.

Guild's Committee for Federal Writers' Publications. *New York Panorama.* New York: Random House, 1938.

Hartog, Hendrik. *Public Property and Private Power: The Corporation of the City of New York in American Law, 1730–1870.* Chapel Hill: University of North Carolina Press, 1983.

Jackson, Anthony. *A Placed Called Home: A History of Low Cost Housing in Manhattan.* Cambridge: MIT Press, 1976.

Select Bibliography

Johnson, Harry and Lightfoot, Frederick S. *Maritime New York in Nineteenth-Century Photographs.* New York: Dover, 1980.

Lockwood, Charles. *Manhattan Moves Uptown.* Boston: Houghton Mifflin, 1976.

Lubove, Roy. *The Progressives and the Slums.* Westport, Conn.: Greenwood Press, 1962.

Luke, Myron H. *The Port of New York 1800–1810.* New York: New York University Press, 1953.

Mann, Roy. *Rivers in the City.* New York: Praeger, 1973.

Mayer, Harold M. and Wade, Richard C. *Chicago: Growth of a Metropolis.* Chicago: University of Chicago Press, 1969.

McKay, Richard C. *South Street: A Maritime History of New York.* New York: Putnam, 1934.

Moehring, Eugene. *Public Works and the Patterns of Urban Real Estate Growth in Manhattan 1835–1894.* New York: Arno Press, 1981.

Morrison, John H. *History of New York Shipyards.* New York: Wm. F. Sametz, 1909.

Moscow, Henry. *The Street Book: An Encyclopedia of Manhattan's Street Names and Their Origins.* New York: Hagstrom Company, 1978.

Rosebrock, Ellen Fletcher. *Walking Around in South Street: Discoveries in New York's Old Shipping District.* New York: South Street Seaport Museum, 1974.

Rush, Thomas E. *The Port of New York.* Garden City: Doubleday, Page, 1920.

Shumway, Floyd M. *Seaport City: New York 1775.* New York: South Street Seaport Museum, 1975.

Stokes, I. N. Phelps. *The Iconography of Manhattan Island 1498–1909.* 6 vols. 1915. New York: Arno Press, 1967; reprint ed.

Sworkis, Martin B., ed. *The Port of New York and the Management of Its Waterfront.* New York University Graduate School of Public Administration, 1959.

Van der Zee, Henri and Barbara. *A Sweet and Alien Land: The Story of Dutch New York.* New York: Viking, 1978.

Whitehill, Walter Muir. *Boston: A Topographical History.* Second ed. Cambridge: The Belknap Press of Harvard University Press, 1982.

Wille, Lois. *Forever Open Clear and Free: The Historic Struggle for Chicago's Lakefront.* Chicago: Henry Regenery, 1972.

Workers of the Writers' Program of the Work Projects Administration in the State of Massachusetts. *Boston Looks Seaward. The Story of the Port 1630–1940.* Boston: Bruce Humphries, 1940.

Wrenn, Douglas M. *Urban Waterfront Development.* Washington, D.C.: Urban Land Institute, 1983.

Writers' Program of the Work Projects Administration for the City of New York. *A Maritime History of New York.* Garden City: Doubleday Doran, 1941.

Waterfront Plans, Reports, and Studies—General

Abt Associates Inc. *Water and the Cities: Contemporary Urban Water Resource and Related Land Planning.* Washington, D.C.: Office of Water Resources Research, U.S. Department of Interior, June 1969.

Battelle Memorial Institute, *Evaluating Urban Core Usage of Waterways and Shorelines.* Washington, D.C.: Office of Water Resources, U.S. Department of the Interior, 1971.

Boston, City of. Boston Redevelopment Authority. *Harborpark. A Framework for Planning Discussion,* 1984.

———. *The City of Boston Municipal Harbor Plan,* 1990.

Chung, Hyung C. *Harbor and Waterfront Development Planning.* Bridgeport, Conn.: Higher Education Center of Urban Studies, March 1977.

Cowey, Ann Breen; Kaye, Robert; O'Conner, Richard; and Rigby, Richard. *Improving Your Waterfront: A Practical Guide.* U.S. Department of Commerce, National Oceanic and Atmospheric Administration, 1980.

Farrell, Virginia. *Development and Regulation of the Urban Waterfront: Boston, San Francisco and Seattle.* Princeton: The Center for Energy and Environmental Studies, Princeton University, 1980.

Harney, Andy Leon, ed. *Reviving the Urban Waterfront.* Washington, D.C.: Partners for Livable Places, n.d.

Heritage Conservation and Recreation Service. Department of the Interior. *Urban Waterfront Revitalization: The Role of Recreation and Heritage.* 2 vols. Washington, D.C.: Department of the Interior, n.d.

Select Bibliography

Moore, Arthur Cotton Associates. *Bright, Breathing Edges of City Life: Planning for Amenity Benefits of Urban Water Resources.* Washington, D.C.: Office of Water Resources Research, U.S. Department of the Interior, Project No. C-2141, 1971.

Wallace, McHarg, Roberts and Todd. *Inner Harbor and Municipal Center Plan.* Baltimore: Department of Planning and the Greater Baltimore Committee, 1964.

Wisconsin, State of. *Waterfront Renewal.* Madison: Wisconsin Department of Resources Development, 1966.

U.S. Department of the Interior. *Urban Waterfronts: Findings and Recommendations.* 2 vols. Washington, D.C., 1979.

Waterfront Plans, Reports, and Studies—New York City

Baiter, Richard. Office of Lower Manhattan Development. Department of City Planning. Battery Park City Authority. *Lower Manhattan Waterfront.* New York, June 1975.

Battery Park City Authority. *Master Development Plan—Battery Park City.* New York: Battery Park City Authority, August 1969.

——. *Battery Park City. Design Guidelines for the North Residential Neighborhood,* 1994.

Battery Park City Authority and Cooper, Alexander Associates. *Battery Park City Draft Summary Report and 1979 Master Plan,* 1979.

Chelsea Piers L.P. *Chelsea Piers Proposal, Recreation and Entertainment Complex,* 1992.

Downtown-Lower Manhattan Assn. Inc. *Lower Manhattan Recommended Land Use, Redevelopment Areas, Traffic Improvements. First Report.* Skidmore, Owings and Merrill Consultants, David Rockefeller, Chairman, 14 October 1958.

Ebasco Services Inc., Moran, Proctor, Mueser & Rutledge; and Eggers & Higgins. *The Port of New York Comprehensive Economic Study for Manhattan North River: Development Plan 1962 to 2000,* 28 November 1962.

Hudson River Park Conservancy. *Hudson River Park Concept and Financial Plan,* 1995.

Lopez, Stephen. *Changing New York City's Waterfront: A Citizen's*

Guide. New York: Cornell University Cooperative Extension, 1980.

Manhattan Borough President Ruth W. Messinger. *Comprehensive Manhattan Waterfront Plan,* 1995.

New York City. Community Board Six Manhattan. *Stuyvesant Cove 197-A Plan,* 1996.

New York City. Department of Parks & Recreation. *Waterfront Management Plan,* 1990.

New York City. Department of Planning. *Amendment to the Zoning Resolution of the City of New York,* August 1993.

———. *New York City Comprehensive Waterfront Plan, Reclaiming the City's Edge,* 1992.

New York City. Department of Planning. Coastal Zone Management. *Draft New York City Local Coastal Zone Management Program,* September 1979.

New York City. Department of Planning. Waterfront Revitalization Program. *Investing in the Waterfront: New York City's Waterfront Revitalization Program,* 1997.

———. *Waterfront Planning Areas and Projects,* February 1984.

———. *Waterfront Revitalization.* New York, 1982.

New York City. Economic Development Corporation and Manhattan Community Board Six. *A New Plan for Stuyvesant Cove,* 1997.

New York City. Planning Commission. *The Manhattan Waterfront: Prospects and Problems,* April 1965.

———. *The Port of New York Proposals for Development,* September 1964.

———. *The Waterfront: Supplement to Plan for New York City.* New York, January 1971.

New York City. Planning Commission. Comprehensive Planning Workshop. *The New York City Waterfront,* June 1974.

New York City. 208 Areawide Waste Treatment Management Planning Program. *Water Quality Management Plan,* April 1979.

New York State. Urban Development Corporation. *Wateredge Development Proposal.* New York, May 1971.

Port Authority of New York and New Jersey. *Targeted Inner Harbor Development,* October 1980.

Quennell Rothschild Associates. *East River Docks Urban Design/Landscape Study,* 1992.

Select Bibliography

Regional Plan Association. *The Lower Hudson: A Report of the Second Regional Plan*. New York, December 1966.

Tri-State Regional Planning Commission. *The Changing Harborfront*. New York, March 1966.

Wallace, McHarg, Roberts and Todd; Wittlesey, Conklin and Rossant; and Frank M. Voorhees and Assoc. *The Lower Manhattan Plan*, 1966.

Wateredges. 2 vols. New York: New York City Bicentennial Corporation, 1975.

West Side Task Force. *Final Report,* 1987.

West Side Waterfront Panel. *A Vision for the Hudson River Waterfront Park,* 1990.

Index

Architects, bridges, buildings, cities, geological and physiographic features, highways, Manhattan, Manhattan Borough Presidents, New York City, New York State, parks, piers by name, railroad companies and lines, streets, and vessels by name are indexed under those entries.

Index

Index

Index

Index

Index

Index

Index